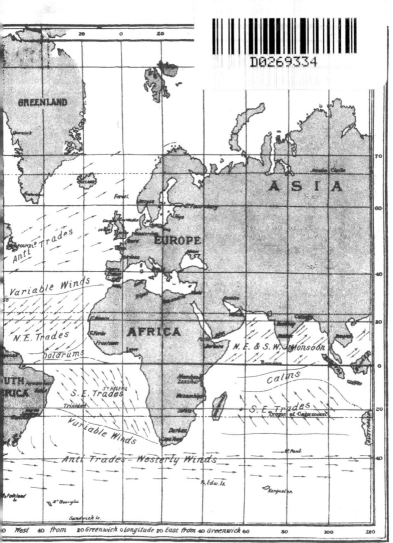

GREENLAND

ASIA

EUROPE

Anti Trades

Variable Winds

N.E. Trades

Doldrums

AFRICA

N.E. & S.W. Monsoon

Calms

S.E. Trades

S.E. Trades

SOUTH AMERICA

Trinidad

Variable Winds

Durban

Cape Town

Anti Trades - Westerly Winds

P. Edw. Is.

Kerguelen

St Paul

Falkland Is.

S. Georgia

Sandwich Is.

West 40 from 20 Greenwich 0 longitude 20 East from 40 Greenwich 60 80 100 120

Scrip. JAMES BROWN & SON, DEL. GLASGOW.

SKYLARKS AND SCUTTLEBUTTS

SKYLARKS AND SCUTTLEBUTTS

~

A Treasure Trove of
Nautical Knowledge

LORENZ SCHRÖTER
Translated by ALAN BANCE
with additional material by DAVID REEVE

~

Granta Books
London

Granta Publications, 2/3 Hanover Yard, Noel Road, London N1 8BE

First published in Great Britain by Granta Books, 2007

Copyright © Lorenz Schröter
Translation copyright © Alan Bance
Additional material © David Reeve

Lorenz Schröter has asserted his moral right
under the Copyright, Designs and Patents Act, 1988,
to be identified as the author of this work.

All rights reserved. No reproduction, copy or transmissions of
this publication may be made without written permission. No
paragraph of this publication may be reproduced, copied or
transmitted save with written permission or in accordance with
the provisions of the Copyright Act 1956 (as amended). Any
person who does any unauthorized act in relation to this
publication may be liable to criminal prosecution and civil claims
for damages.

A CIP catalogue record for this book
is available from the British Library.

1 3 5 7 9 10 8 6 4 2

ISBN 978-1-86207-974-8

Typeset by M Rules

Printed and bound in Great Britain by
Mackays of Chatham plc

SKYLARKING is the act of running about the rigging of a vessel in sport.

A SCUTTLEBUTT was a water cask on a ship. Since sailors exchanged gossip when they gathered at a scuttlebutt to drink, it became a familiar term for gossip or rumours, or someone who purveyed them.

Volume of Water

~

The total volume of water on the Earth is
1,385,984,610 km^3; it is made up as follows:

Seawater 1,338,000,000 (96.5%)
Glaciers and snow 24,064,000 (1.74%)
Groundwater 23,400,000 (1.7%)
Permafrost 300,000 (0.022%)
Lakes and ponds 176,400 (0.013%)
Ground humidity 16,500 (0.001%)
Fog and clouds 12,900 (0.001%)
Marshes and swamps 11,470 (0.0008%)
Rivers 2,120 (0.0002%)
Water in living creatures . . . 1,120 (0.0001%)

Coastlines

~

(Source: CIA World Factbook)

Monaco . . . . 4.1 km
Bosnia and
   Herzegovina . . . . . . 20 km
Jordan . . . . 26 km
Congo . . . . 37 km
Palestine . . . . 40 km
Slovenia . . . . 46.6 km
Togo . . . . 56 km

Iraq . . . . 58 km
Belgium . . . . 66.5 km
USA . . . . 19,924 km
Australia . . . . 25,760 km
Japan . . . . 29,751 km
Indonesia . . . . 54,716 km
Canada . . . . 202,080 km
(because of its islands)

Why the British coastline is Infinitely long

Why is it, wondered the mathematician Benoît Mandelbrot, that every atlas gives a different figure for the length of the British coastline? He discovered that the coastline became longer according to the level of detail in the map being used. The more detailed the map, the smaller the coastal indentations it had to include, and these small additions added up to ever larger totals. Mandelbrot concluded that if you took into account every grain of sand on the seashore, every 'dry' molecule that had just been separated off from the nearest 'wet' molecule, the length of the coastline would be infinite.

When it comes to coastlines, mountains and clouds, nature deviates from the standard Euclidean shapes of cone, cube and sphere. In his epoch-making book of 1975, *The Fractal Geometry of Nature*, Mandelbrot demonstrated that shapes in nature are not smooth or regular, but jagged and (seemingly) irregular.

A mountain, writes the French-Polish mathematician, never completely fills a three-dimensional, cube-shaped space. The dimensions of a mountain can therefore never be expressed as a whole number, but only as a fractal (Latin *fractus* = 'broken'), a value between two and three. The same applies to a coastline, which would be purely one-dimensional if it were a simple line. But in fact, as a measuring grid will show, the line incorporates surface areas. By Benoît Mandelbrot's calculation, therefore, the British shoreline has a fractal dimension of 1.25.

Today, fractals are used to work out share prices and weather forecasts, in programming washing machines and setting up e-mail servers. They are applied, in fact, to every area where both order and chance prevail: in other words, wherever chaos is apparent.

Total quantity of foam currently on the world's seas

720,000 km^2

The Biggest Islands

Greenland ...... 2,175,600 km^2
New Guinea ...... 790,000 km^2
Borneo .... 743,330 km^2
Madagascar .... 587,041 km^2

States located on artificial islands

Atlantis, also called the Isle of Gold, lay south-west of Florida and began the history of mini-states sited on platforms. The firm of *Acme General Contractors, Inc.*, proprietor Louis Ray, intended to conduct offshore banking and other financial services from here, beyond the reach of national jurisdictions, but Atlantis was destroyed by a storm in 1964. A new attempt under the name of *Grand Capri* failed in 1966.

New Atlantis was founded on 4 July 1964 on a tiny platform in the Caribbean, with a population of eight citizens. The local currency consisted of fish hooks, sharks' teeth, and carob pods. But the islanders were careless about security: passing fishermen helped themselves to wood from the construction, and the young nation was tipped into the sea. The state's founder, Leicester Hemingway, a brother of the writer, tried his luck again on the nearby artificial island of *Tierra del Mar*, but this second venture, too, ended in failure. For the next five years Hemingway edited the anglers' magazine *The Bimini Out Islands News*. He developed severe diabetes, and in 1982 committed suicide – as did his father Clarence, his brother Ernest, his sister Ursula, and his great-niece Margaux. Today his daughter Hilary, who grew up on New Atlantis, is an expert on abduction by UFOs.

The Republic of Rose Island: On 1 May 1968 the engineer Giorgio Rosa declared independence for his 1200-square-metre platform in the Mediterranean off Rimini, and introduced Esperanto as the official language of his state. Two months later the little island was (illegally) destroyed by the Italian navy.

Abalonia: The plan was to anchor the concrete ship SS *Abalone* as a floating state in the rich fishing grounds of Cortes Bank off California. But she sank on her maiden voyage in 1969, and her crew of anarchists and pacifists narrowly escaped drowning. Today the wreck is a favourite with divers in the know.

Sealand: This Second World War radar platform, measuring 47 by 13 metres, is sited in the North Sea at 51° 53' 40" N, 1° 28' 57" E, six miles off the Suffolk coast. In September 1967 it was occupied by the radio pirates Roy Bates and Ronan O'Rohilly. After a disagreement between the two adventurers, there was a short but fierce battle involving guns, Molotov cocktails and a flame-thrower. Bates emerged victorious, and from then on declared himself 'Prince Roy the First', ruling Sealand as an independent principality. The British navy wanted to put an end to the whole affair, and arrested Bates. However, it was determined in court that as Sealand lay outside the three-mile zone it was beyond British jurisdiction.

A few years later there was the first attempt at a coup against the 611-square-metre mini-state. A German called Alexander Aschenbach, who had been appointed by Bates as Sealand's Foreign Minister and Premier for life, hired a couple of Dutch mercenaries and kidnapped his sovereign ruler's son, Crown Prince Michael. Roy Bates lost no time in renting a helicopter, wresting back his principality, and convicting the usurper of high treason. The Dutch and German authorities urged the British to put an end to this farce, but they took no action. Today, Aschenbach leads a government-in-exile from Germany.

During the Falklands war Argentina tried to rent Sealand as a rocket-launching site, clearly a step too far for Bates. To prevent the creation of other mini-states, the British government has now extended its sovereignty to twenty miles, and blown up some of the old platforms. Undeterred, Prince Roy signed a leasing agreement with an enterprise called Haven Co. Ltd, which set up an internet server on Sealand. The firm reckoned that on Sealand computers would be beyond the reach of governments.

Fastest Glaciers

(distance travelled per year according to NASA measurements from space)
Kangerdlugssuaq Glacier, Greenland ≈ 14 km
Jakobshavn Isbrae, Greenland ≈ 12.6 km
Thwaites Glacier, Amundsen Sea, Antarctic ≈ 3 km
Pine Island Glacier in Bellingshausen Sea, Antarctic ≈ 2 km

Channel Swimmers

In 1815, after the defeat of Napoleon, an Italian soldier called *Jean-Marie Saletti* jumped overboard from a British ship and swam back to the French coast, where he was picked up by a French fishing boat. As he had not actually swum from shore to shore, the venerable Channel Swimming Association does not include him in its records. Officially, the history of Channel swimming began when *Captain Matthew Webb* stepped into the water on 24 August 1875 near Dover and set off at twenty strokes a minute. His support boat supplied him only with beer, brandy and tea. He was carried by the tide on a long S-shaped course until finally, after covering more than 70 kilometres in the water, Captain Webb sighted the coast of France. By now exhausted and reduced to twelve strokes a minute, he had to battle for a further nine hours against the ebb tide, and finally staggered ashore after 21 hours and 45 minutes in the water.

Despite 71 further attempts in the next 26 years, nobody

managed to repeat Captain Webb's achievement during his lifetime. One man, *Jabez Wolfe*, made the attempt no fewer than 20 times, and on one occasion was forced to give up only a couple of kilometres from land. In 1911 *Thomas Burgess* became the first since Webb to complete the crossing, at his twentieth attempt.

Captain Webb became rich and famous, but there was no happy ending. He gambled away his entire fortune in casinos, and at 35 he took on the ultimate challenge – Niagara Falls, equally popular with honeymooners and suicides. Clad only in his swimming costume, the most famous swimmer of his day jumped into the river. It was not until four days later that his battered body was washed up on shore. Today Captain Webb's moustachioed statue stares down at his would-be successors in Dover Harbour. The same conditions apply to them as to the pioneer Channel swimmer: no wetsuits or flippers. Since Saletti and Webb, 725 people have swum the distance of – as the crow flies – 32 kilometres. The fastest man one way across the Channel is Chad Hundeby (USA), 7 hours 17 minutes in 1994, swimming England to France. The fastest man swimming France to England is Richard Davey (England), 8 hours 5 minutes in 1988. The latest successful Channel swimmers recorded by the Channel Swimming Association on 31 December 2006 are Sønnøve Cirotski (Norway, 14 hours 43 minutes) and Erica Moffett (USA, 14 hours 19 minutes), both in 2006.

Olympic medals won by South Sea Islanders

Paea
Wolfgramm
from the Tonga Islands
won a silver medal for super-
heavyweight boxing
in Atlanta in
1996

Diving Records (humans)

10,916 m ≈ distance below sea level dived by Jacques Piccard on 23 January 1960 in his bathyscaphe, *Trieste*. He reached what is believed to be the deepest point on the earth's surface, the Challenger Deep, in the Mariana Trench.

313 m ≈ dived by Mark Ellyat using compressed air on 18 December 2003.

162 m ≈ dived by Loïc Leferme on 20 October 2002 and Pipin Ferraras on 18 January 2000, without the aid of compressed air.

8:06 mins ≈ length of time Martin Stepanek held his breath under water on 3 July 2001.

120 hours ≈ length of time Jerry Hall spent swimming under water using compressed air on 29 August 2004.

Diving Records (animals)

~

Animal	Duration	Depth
Sealion	10 mins	274 m
King penguin	20 mins	535 m
Leatherback sea turtle	120 mins	750 m
Bottlenose whale	70 mins	1453 m
Sea elephant	90 mins	1581 m
Sperm whale	73 mins	2035 m

Number of Fatal Shark Attacks

~

136	Australia (recorded since 1700)
54	USA (since 1670)
41	South Africa (since 1905)
25	Papua New Guinea (since 1925)
20	Brazil (since 1930)
11	Mascarene Islands (since 1828)
11	Hong Kong (since 1580)
9	Cuba (since 1749)
9	New Zealand (since 1825)
8	Iran (since 1580)
8	Japan (since 1580)
8	Greece (since 1847)
8	Fiji (since 1925)

Uses for sharkskin

~

The Romans covered their helmets with it. The ancient Germanic tribes used to wrap it round their sword-hilts for a better grip. In the South Pacific it is used as emery paper. The application to aeroplane surfaces of 'riblet film', inspired by sharkskin, improves drag reduction and produces a fuel saving of 3%. An anti-fouling paint for ships' hulls was developed which emulated the repellent properties of sharkskin.

How to prepare a sharkskin to cover a drum

Carefully remove the skin from the flesh of the shark with an ordinary dinner knife. Then keep the skin for two days in a preserving jar filled with urine. Wash with seawater to get rid of the smell. Squeeze out any moisture, and then stretch the skin across a cardboard box, fastening it with drawing pins. Next, scrape the skin with a sharp knife to make it thinner. You now have a piece of *shagreen*, which makes a drum skin for a *puniu* or Hawaiian coconut knee drum.

Watson and the Shark

The painting *Watson and the Shark* by John Singleton Copley (1738–1815) was inspired by an event that took place in Havana, Cuba, in 1749. Brook Watson (1735–1807), a fourteen-year-old orphan serving as a crew member on a trading ship, was attacked by a shark while swimming alone in the harbour. Despite the efforts of his valiant shipmates, Brook Watson was repeatedly attacked by the shark, which bit off his right foot. His right leg was later amputated below the knee. Watson eventually became a successful London merchant, and a chance meeting with the artist John Copley in the summer of 1774 led to a commission to re-create the ghastly scene. In April 1798 the painting was exhibited at the Royal Academy in London, causing a sensation. Newspapers carried the story in gruesome detail.

Watson went from strength to strength and became Lord Mayor of London 1796–7 and Sir Brook Watson 1st Baronet in 1803. For his coat of arms he requested the inclusion of his missing right leg in the upper left corner of the shield and Neptune brandishing a trident to ward off an attacking shark.

The painting was owned by Watson until his death in 1807 and bequeathed by him to Christ's Hospital Boys' School. He hoped his personal triumph over adversity would be a 'most useful lesson to youth'. In 1963 the picture was sold to the National Gallery of Art, Washington, DC, though Christ's Hospital retain a copy.

How to tell the age of a fish

All fish with bones develop little auditory ossicles, otoliths or 'earstones', where concentric accretions of calcium carbonate are built up, comparable with the growth rings of a tree. Every layer represents a year. Take out the pair of otoliths and put them in a bowl of water. (With flatfish, you should only examine the otolith from the 'blind' underside.) Dissect the material with a scalpel, and count the rings. Sometimes the otolith has to be heated with an alcohol burner. Be careful: as it becomes hot and gives off smoke, do not let it get more than slightly singed. Count all the rings you can see, including half-rings, and compare the result with that obtained from the other otolith.

How to artificially inseminate a sea urchin

You will need:
≈ about four sea urchins (to increase your chances of having at least one male and one female)
≈ 0.5 molar solution of potassium chloride
≈ seawater, strained through a coffee filter
≈ beakers
≈ hypodermic needle
≈ pipettes
≈ microscope slide with cover-glass
≈ Petri dishes
≈ microscope

What to do:
Place the sea urchin flat side down in a glass filled with sea water warmed to about 18 to 20 degrees centigrade. Drop some potassium chloride in the mouth area, and the sea urchin should release its eggs or sperm. The white threads are sperm, while the ova float in a cloud to the bottom. If nothing has happened after five minutes, pick up the sea urchin, shake it, and put it back in the water. If this does not produce a result, inject it with a few drops of potassium chloride. Insert the needle in the soft area near the 'Aristotle's lantern', the mouth aperture, and inject the potassium chloride into the inner side, i.e. the gonad area. If this does not work, select another sea urchin.

Place the ova on a microscope slide, add sperm, and observe the result. A membrane should soon form around the ovum: a new sea urchin is in the process of formation. There are varieties, like the red sea urchin (*Strongylocentrotus franciscanus*), that live for 200 years, while remaining capable of breeding. In fact, 100-year-old sea urchins produce sperm and ova that are better in quantity and quality than those from younger ones.

Picasso and the sea urchins

~

Skull, sea urchin, and lump on a table (1946)
Skull and three sea urchins (1947)
Woman with sea urchins (1946)
Man eating sea urchin (1946)

The sea urchin as a symbol

For Dalí (*Un chien andalou*) it represents the
female genitals (minus spines).

In Hieronymus Bosch (*The Wedding Feast at
Canaan*) it is the philosopher's stone.

For the Celts (*dragon's egg*) it symbolized the
endless, hidden continuity of life.

The greatest wartime shipping disasters

Wilhelm Gustloff, 30 January 1945: 9,000 refugees died.
Tango Maru and *Ryusei Maru*, 29 June 1944: 8,000 forced
  labourers and soldiers lost their lives in a bombardment.
Cap Arcona, 3 May 1945: 7,000–8,000 former prisoners of a concen-
  tration camp were killed in an attack by British fighter bombers.
Goya, 16 April 1945: 7,000 refugee casualties.
Junyo Maru, 18 September 1944: 5,620 forced labourers and
  prisoners of war lost.
Toyama Maru, 29 June 1944: 5,600 soldiers died.
Steuben, 9 February 1945: 4,000 refugees perished.
Lancastria, 17 June 1940: 2,000 deaths, both civilian and military.
Arisan Maru, 24 October 1944: loss of 1,792 prisoners of war.
Koshu Maru, 3 August 1944: 1,540 forced labourers died.
Tsushima Maru, 22 August 1944: 1,529 civilian casualties.

The greatest peacetime shipping disasters

The ferry *Doña Paz*, 20 December 1987: 4,317 passengers lost.
The ferry *Joola*, 26 September 2002: loss of 1,863 passengers.
Sultana, 27 April 1865: 1,700 passengers died.

The sailing junk *Tek Sing*, January 1822: more than 1,600 passengers and crew drowned.

Titanic, 14 April 1912: 1,504 passengers and crew lost.

Europeans in Australia

1606   Willem Janszoon, captain of the *Duyfken*, was the first European to make landfall on the new continent (Cape York).

1606   Louis Vaez De Torres negotiated the Torres Strait with the *San Pedro* and the *Los Tres Reyes*.

1619   Frederik de Houtman and Jakob Dedel landed from the *Dordrecht* and the *Amsterdam* at the mouth of the Swan River, and later on the Houtman Albrolhos Islands and Rottnest Island.

1622   The *'t Wapen van Hoorn* ran aground in Shark Bay, but was re-floated on the tide. In the same year the crew of the *Leeuwi* mapped the territory at the most south-westerly point of Australia.

1622   The *Tryall* foundered on the reef now known as Tryal Rocks in Western Australia. Forty-six seamen, including Captain John Brookes, got ashore in two boats. Ninety-three men died.

1623   J. Carstenszoon and W. van Coolsteerdt landed from the *Pera* and the *Arnhem* at Cape York. In the same year Claus Hermanszoon landed the *Leijden* south of Dirck Hartog Island, and Australia's first European baby was born there.

1629   The *Batavia*, commanded by Fransisco Pelsaert, was shipwrecked on the Houtman Albrolhos Islands, and the crew were engaged in fighting with the aboriginals. A mutiny followed. The *Sardam* rescued the stranded men, but two mutineers, Wouter Loos and Jan Pelgrom de Bye, were left marooned ashore. Thus began the history of Australia as a land of banishment.

1642   Abel Janszoon Tasman conquered what is now Tasmania. With his ships *Limmen*, *Zeemeeuw* and *Braq* he was the first to circumnavigate the whole continent.

1656   68 sailors on the *Vergulde Draeck* were shipwrecked on a reef near Perth. Nothing more was ever heard of them. The site of the wreck was discovered in April 1963 by a group of spear fishermen.

1658   Looking for survivors, Samuel Volckersen landed on Rottnest Island and on the mainland. An artist on board made the first drawings of aboriginals.

1688   Captain Read and William Dampier anchored their ship *Cygnet* off the coast of Western Australia for two months to carry out repairs.

1712   The *Zuytdorp* was wrecked 60 miles north of the mouth of the Murchison River. The site of the sailors' encampment was discovered in 1927.

1727   After the shipwreck of the *Zeewijk*, Jan Steyn was the first to build a European ship in Australia, the *Sloepi*; he sailed in it to Indonesia.

1768   Louis Antoine de Bougainville, commander of the *Boudeuse* and the *Étoile*, was thwarted by the Great Barrier Reef.

1770   James Cook was the first European to go ashore on the east coast of Australia; his ship later ran aground on the Great Barrier Reef.

Pioneers of Transatlantic Flight

(Principal pilots' names are given in full; crew members are
listed by surname only.)

May 1919: Albert Read, Stone, Hinton, Rodd, Rhodes, Howard,
Breese. One stop in the Azores.

June 1919: John Alcock, Arthur Whitton Brown. Newfoundland
to Ireland direct.

July 1919: George Herbert Scott, Angus, Browdie, Burgess, Cooke,
Cross, Durrant, Edwards, Evenden, Forteath, Gent, Graham,
Gray, Greenland, Harris, Luck, Maitland, Mayes, Mort,
Northeast, Parker, Powell, Pritchard, Ripley, Robinson, Scull,
Shotter, Smith, Thirlwall, Turner, Albert Watson, Lansdowne,
Hensley, Ballantyne (stowaway). Scotland to New York direct.

August 1924: Lowell Smith, Arnold, Nelson, Harding; England to
Labrador via Iceland.

October 1924: Hugo Eckener, Auer, Belscr, Christ, Fischer,
Fleming, Freund, Grofzinger, Kiefer, Knorr, Ladwig, Lang,
Lehmann, Leichtle, Martin, Marx, Pabst, Praff, Pruss, Sammt,
Scherz, von Schiller, Schwendt, Siegle, Specy, Tassler,
Tielmann, Wittemann, Steele, Klein, Kraus, Kennedy.
Zeppelin (airship) flight from Friedrichshafen in Germany
direct to New Jersey (a trip 650 kilometres longer than
Lindbergh's).

January 1926: Ramón Franco (brother of the dictator), de Alda, Rada. By seaplane from Spain to Brazil, stopping en route.

March 1927: Sarmento de Beires, Jorge de Castilho, Gouveia. By seaplane from Portugal to Brazil, with stops.

May 1927: Charles Lindbergh. New York to Paris, solo non-stop flight. He took off from Long Island, not from the mainland.

Canals

Name	Date opened	Length in km
St Lawrence Seaway	1959	304
White Sea–Baltic Canal	1933	227
Suez Canal	1869	163
Moscow Canal	1937	128
Volga–Don Canal	1952	101
Kiel Canal	1895	98.7
Houston Ship Canal	1940	91.2
Alphonse XIII Canal, Spain	1926	85
Panama Canal	1914	86

Panama and Suez

Panama: The lock chambers are 305 m long and 33.5 m wide. The canal is 86 km long with a maximum depth of around 13.5 m. A passage takes 8 hours. At present the canal has 2 lanes but a 3rd lane and increased lock chamber size are planned to accommodate larger container ships.

Suez: The canal is 163 km long and 80–135 m wide with no lock chambers. Most of the canal has a single traffic lane with several passing bays. It is planned to increase the depth of the canal before 2010 in order to allow passage of the very largest container ships.

Channel Tunnel

Tunnel opened	*1994 (1 service + 2 train tunnels)*
Eurostar passengers 2005	8.2 million
Tunnel length	50 km (39 km under the Channel)
Tunnel depth	40 m average below Channel bed
Construction Workers	13,000
Geology	Mostly Cretaceous chalk marl
Rubble volume	3 times the Pyramid of Cheops
Train speed	Up to 160 km per hour
Folkestone to Calais	35 minutes
London to Paris	2 hours 15 minutes
Construction cost	£9.8 billion (80% overrun)

Monogamous fish

Clownfish
Sea horses
Kribensis (*Pelvicachromis pulcher*)

Marine hermaphrodites

~

Butterfly fish ≈ Anemones
Several species of sponge and coral
Zander ≈ Nassau grouper

Eels (*inclining to the female sex at first, they pass through a hermaphroditic stage before adopting their ultimate sexual identity*)

Painted comber (*can reproduce up to six times in a few hours, alternating in sex*)

Coral-fish (*live together in groups of five; the largest fish is the only female in the group. If the female dies, the male next in line takes her place – and changes into a female.*)

Fins and Flippers

~

Fish Have Vertical Fins.

Mammals (e.g whales) Have Horizontal Flippers.

Finding Atlantis

~

Places where it is claimed to have been found:
Helike, Peloponnese ≈ Madeira, Portugal ≈ Thera (present-day Santorini), Aegean ≈ Tartessos, Spain ≈ Chott el Djerid, Tunisia ≈ the mouth of the Elbe ≈ America ≈ Antarctic ≈ Troy ≈ Black Sea ≈ under the Sahara ≈ Heligoland ≈ Ireland ≈ Nigeria

Sea cucumbers and us

Chinese palaeontologists reported recently in the scientific journal *Nature* that humans are closely related to sea cucumbers (sea slugs), sea urchins and starfish. Researchers discovered a fossil proving the link between echinoderms ('spiny' forms) and vertebrates.

Squid will rule the world!

The warming of the oceans has brought them victory: the biomass, i.e. the total body weight of all individual specimens added together of cephalopods (more familiarly known as octopods, squid and octopuses), now exceeds that of humanity as a whole. The previous front-runners, ants and plankton, will have to start trying harder …

The price of an intelligent kraken

The National Resource Center for Cephalopods in Galveston, Texas, has measured the intelligence of cephalopods and found they are the cleverest of the invertebrates, and brighter than the average dog. These are typical prices for squid (plus postage and packing):

Species	Embryo	>1 g >30 mm	1–50 g >60 mm	50–300 g >150 mm	300–1000 g >150 mm
Euprymna scolopes	$10	$30	$45		
Sepioteuthis lessoniana	$15	$30	$50	$70	$90
Sepia officinalis	$5	$25	$35	$45	$50
Sepia pharaonis	$15	$30	$45	$60	$80
Octopus bimaculoides	$10	$30	$40	$50	$60

Female pirates

Ch'iao K'uo Fü Jën: Chinese, *c.* 600 BC
Queen Artemisia of Halicarnossus: Greek, *c.* 480 BC
Elissa ('Dido'): founder of Carthage, *c.* 470 BC
Queen Teuta of Illyria: c. 230 BC
Princess Sela: Viking, *c.* AD 420
Wigbiorg, Hetha, Wisna: Vikings, c. 800.
Alfhild (or *Alwida*): Viking, *c.*850.
Ladgerda: Viking, *c.*870.
Grace (Gaelic: *Granuaile/Grainne*) *O'Malley:* Irish, *c.*1500.
Lady Killigrew: Irish, 1530–70.
Mrs Peter Lambert of Aldeburgh: English, late 16th century
*Elizabeth Patrickson: c.*1634
Jacquotte Delahaye: Caribbean pirate, *c.*1650
Anne Dieu-le-veut: Caribbean pirate, *c.*1660
La Marquise de Frèsne: made the Mediterranean unsafe in the
    late 17th century
Anne Bonny: born 1700 in Ireland, supposedly executed in the
    Caribbean in 1720
Mary Read: born 1690 in London, partner of Anne Bonny, died
    *c.*1720 in the Caribbean
Mary Harvey (alias *Harley* or *Mary Farlee*): American, *c.*1725
Rachel Wall: American, *c.*1780
'Sadie the Goat': American, *c.*1860
Catherine Hagerty: Australian, *c.*1806
Margaret Jordan: Canadian, *c.*1809
Cheng I Sao (or *Cheng Yih Saou*): Chinese, commanded *c.*1810
    about 800 junks and 1,000 boats, with crews of 70,000 to
    80,000 men and women in total
Gertrude Imogene Stubbs, alias 'Gunpowder Gertie': terrorized
    the lakes and rivers of British Columbia, Canada, 1898–1903
Lo Hon-cho (or *Honcho Lo*): after her husband's death in 1921,

took command of a pirate fleet of 64 junks, and supported the Chinese Revolution

Wong: with her 50 boats, she joined Lo Hon-cho in 1922

Lai Sho Sz'en (or *Lai Choi San*): commanded 12 junks, 1922–39

P'en Ch'ih Ch'iko: commanded about 100 pirates, 1936

Huang P'ei-mei: Chinese; led 50,000 pirates between 1937 and 1950

Linda: Filipina, *c.*1980

Land-locked nations

Afghanistan ≈ Andorra ≈ Armenia ≈ Austria ≈ Azerbaijan ≈ Belarus ≈ Bhutan ≈ Bolivia ≈ Botswana ≈ Burkina Faso ≈ Burundi ≈ Ethiopia ≈ Hungary ≈ Kazakhstan ≈ Kyrgyzstan ≈ Laos ≈ Lesotho ≈ Liechtenstein ≈ Luxembourg ≈ Malawi ≈ Mali ≈ Moldavia ≈ Mongolia ≈ Nepal ≈ Niger ≈ Paraguay ≈ Republic of Chad ≈ Republic of Macedonia ≈ Rwanda ≈ San Marino ≈ Slovakia · Swaziland ≈ Switzerland ≈ Tajikistan ≈ Turkmenistan ≈ Uganda ≈ Uzbekistan ≈ Vatican ≈ Zambia ≈ Zimbabwe

The Mongolian navy

Mongolia is the largest country with no access to the sea. Its capital, Ulan Bator is about 1,500 kilometres from the nearest coast. In 1938 a tugboat was dismantled and transported across the steppes to the shores of Lake Hovsgol, the largest lake in the country. The *Suhkbaatar* currently has a seven-man crew.

Under the Mongolian flag

In the 1980s, a Mongolian student called Ganbaatar won a scholarship to study in Russia. He was assigned by mistake not to Course 1013 (fish-breeding) but to Course 1012 (deep-sea fish). He did his military service in the Mongolian navy, and later put together the Mongolian maritime law code. In February 2003 the Mongolian shipping registry opened its books: it transpired that there were already 300 ships flying under Mongolian colours.

But there were criminal intentions underlying this apparent success. The management team from Singapore hired by the Mongolian government had previously organized the North Korean shipping registry, whose ships were notorious for their frequent involvement in drug smuggling (the regime was short of money). Once the reputation of North Korean ships had been ruined, the shipping company changed its name and nationality – but the ships and the management remained the same. The ship owner, a certain Chong Koy Sen from Singapore, was allowed to register his fleet under the flag of Cambodia. When these colours, too, became associated with cocaine smuggling, a new flag was soon being hoisted on his vessels – that of Mongolia.

The profits of the slave trade: the schooner *La Fortuna*, 1827

Expenses of a voyage from West Africa to Havana:

Purchase price of ship	$3,700.00
Outfitting (sails, cordage)	$2,500.00
Victuals	$ 1,115.00
Price of 220 slaves	$10,900.00
Pay and fees	$19,755.00
Dealers' commissions	$987.75
Bounty and pay on return	$6,410.46
Fees, commissions and $2-sets of clothes for 217 slaves ..	$12, 608.00
Total outlay on voyage	$39,761.21
Sale of slaves	$77,649.00
Auctioning of ship	$3,950.00
Total proceeds	$81,419.00

A net profit of $41,657.79, representing a return of over 100% in just four months

Ransom demanded by Mediterranean pirates for a Christian seafarer *c.*1720

Captain	1,000 Reichstaler
	(1 Reichstaler then = approx. 4s.6d)
Bosun or carpenter	700 Reichstaler
Seaman	60 Reichstaler

Royal Navy Ships 1918 – 2007

Year	Carriers	Battleships	Cruisers	Frigates/destroyers
1918	4	70	143	443
1939	4	9	21	82
1945	11	14	62	801
1950	12	0	29	280
1960	8	2*	14	156
1970	3	4*	3	81
1980	2	2*	0	64
1990	3	2*	0	51
1997	3	3*	0	37
2004	3	3*	0	25
2007	2	2?	0	19

* = large amphibious craft

Year	Submarines	Coastal
1918	147	—
1939	21	—
1945	131	1,319
1950	66	66
1960	54	207
1970	39	54
1980	44	52
1990	29	54
1997	22	?
2004	13	49
2007	13	?

Characteristics of Inuit music

Complex drum rhythms, nasal singing, shrill female voices.
Subjects of songs include hunting, chamber pots, and
particularly notable farts.
Other popular topics are: bad women and good mothers, e.g.
the story of the duck who led her brood safely out
into open water.
Since there is a taboo against mentioning the dead by
name, or any concept associated with them, many
words are substituted, until the meaning is lost
on everybody.

Ten figureheads at Valhalla on Tresco Island, Scillies

~

Salmon, wooden schooner, 178 tons	Built Quebec 1859 Date of Loss 15 January 1871	Fish
River Lune, iron barque, 1,172 tons	Built Wallsend 1868 Date of Loss 27 July 1879	Female
Bosphorus, wooden schooner, 199 tons	Built St Mary's (Scilly) 1840 Date of Loss 29 October 1880	Male, Turk's head
Friar Tuck, wooden tea clipper, 662 tons*	Built Aberdeen 1856 Date of Loss 2 December 1863	Male, Friar Tuck
Colossus, wooden, 3rd rate ship of the line, 1,717 tons	Built Gravesend 1787 Date of Loss 10 December 1798	Decorated stem board
Golden Eagle, French	Date of Loss before 1850	Golden eagle with snake in beak
Palinurus, wooden barque, 300 tons	Built Whitby 1833 Loss Date 27 December1848	Sailor in Formal Dress
Mary Hay, wooden barque, 258 tons	Built Peterhead 1837 Loss Date 13 April 1852	Elegant female half-figure
Volunteer, wooden ketch, 65 tons	Built Aberystwyth 1861 Loss Date 26 January 1911	Figure of soldier with gun
Chieftain, no records	Built ? Loss Date 1856	Male figure in Highland dress

* The Chinese geese on Tresco today are the descendants of those which came ashore from the wreck of the tea-clipper *Friar Tuck.*

The highest peaks on ...

~

Faroe Islands ≈
Slættaratindur, 882m.
Kiribati ≈ Joe's Hill, 13 m.
England ≈ Scafell Pike, 978 m
(has reportedly been scaled by two-year-olds)
Australia ≈ on the mainland, Mount Kosciusko, 2228 m,
or Mawson
Peak, 2745 m, on Heard Island (two-thirds of the way
from Madagascar to the Antarctic)
Denmark ≈ Yding Skovhøj, 173 m
São Tomé & Príncipe ≈ Pico de Sao Tomé, 2024 m.
Indonesia ≈ Mount Carstensz (aka Puncak Jaya), 4884 m
(one of the 'Seven Summits',
the highest peaks on each of the seven continents)
Antarctica ≈ Vinson Massif, 5140 m (also one of the 'Seven
Summits')

Horns and whistles

~

The International Regulations for Preventing Collisions at Sea specify the technical requirements for sound signal appliances on marine vessels. Frequency range and minimum decibel level output is specified for each class of ship, determined by the vessel's length.

250 – 700 Hz 130dB minimum for a ship of less than 75 m	130 – 350 Hz 138dB minimum for a ship of 75 – 200m	70 – 200 Hz 143dB minimum for a ship of more than 200 m in length

What Columbus brought back from America

Pineapples ≈ blueberries ≈ beans ≈ cashew nuts ≈ rubber trees ≈ peanuts ≈ cocoa ≈ potatoes ≈ pumpkins ≈ maize ≈ cassava ≈ pecan nuts ≈ paprika ≈ tobacco ≈ tomatoes ≈ vanilla
Plus turkeys and syphilis.

(Well, OK: since signs of syphilis were found in a few 14th century English monks, the theory that it was brought back by Columbus has somewhat gone out of fashion. But it is indisputable that this venereal disease became a Europe-wide scourge from 1495 onwards, clearly put into circulation by Columbus's crew. The pious brothers must have had a much less contagious variant of the disease.)

What to call a 'big ship'

Handymax = 60,000 dwt *, corresponding to 2,000 TEU**.
*Panamax**** or *panmax* = 80,000 dw/3,000 TEU.
Suezmax = 200,000 dwt/12,000 TEU, 500 metres in length, width 32.31 metres.
VLCC (Very Large Crude Carrier, or tanker) = 320,000 dwt, 336 metres long, 70 metres wide.
ULCC (Ultra Large Crude Carrier, or supertanker) = 550,000 dwt, 334 metres in length, width 45.60 metres.
Malacca-Max = 18,000 TEU, 400 metres long, 60 metres wide, draught 21 metres.

   * dwt stands for 'deadweight' indicating maximum load capacity, adding together the weight of freight, fuel, water, and provisions.
  ** TEU (Twenty Foot Equivalent Unit) is a unit of calculation based on the standard twenty-foot container.
*** Panamaxes are just able to pass through the Panama Canal,

and Suezmaxes through the Suez Canal. Incidentally, there are still no ports available for the planned ULCCs and the 400-metre-long Malacca-Max, which can squeeze through the Straits of Malacca. Just.

The biggest ships of all time

France II (sailing ship) = sail area 6,350 square metres; length 149.50 metres; width 16.90 metres; draught 8.50 metres. Ran aground 1922.

Royal Clipper (sailing cruise ship) = 227 passengers; sail area 5,200 square metres; length 134 metres; width 16.50 metres; draught 5.60 metres.

Taifun (the submarine featured in the film *Red October*) = 175 metres long; 22.80 metres wide. Broken up in 2005.

USS *Nimitz* (aircraft carrier) = length 332.80 metres; width 40.80 metres (flight deck 76.80 metres); draught 11.30 metres.

Cosco Guangzhou (container ship) = 9,500 TEU; length 350.56 metres; width 42.80 metres; draught 14.50 metres.

Pierre Guillaume (tanker) = 553,622 dwt; length 414 13 metres, width 63 05 metres; draught 28.60 metres. Scrapped in 1983.

Knock Nevis, formerly *Seawise Giant, Happy Giant* and *Jahre Viking* (tanker) = 564,763 dwt; length 458.40 metres; width 68.80 metres; draught 24.50 metres. Because of tighter safety regulations, no longer permitted to go to sea; now anchored off Qatar as transit storage for crude oil.

Queen Mary 2 (passenger ship) = 2,620 passengers; length 351 metres; width 59 metres; draught 10 metres.

Freedom of the Seas (passenger ship) = 4,370 passengers; length 339 metres; width 56 metres; draught 8.50 metres.

Freedom (projected passenger ship) = 50,000 passengers; crew of 15,000; length 1,500 metres. Intended to cruise permanently as a mobile tax haven.

Queen Victoria Cunard ocean liner

~

Builder	Fincantieri – Cantieri Navali SpA
	Marghera shipyard near Venice
Gross tonnage	90,000 tons
Length	294 metres
Beam	32.3 metres
Draught	7.9 metres
Height	54.5 metres
Passenger decks	12
Passengers	2,014
Passenger accommodation	1,007 staterooms
Float out	15 January 2007
First master	Captain Paul Wright
Maiden voyage	December 2007 to N. European cities
First 'meeting' of *Queen Victoria*, *Queen Mary 2* & *Queeen Elizabeth* 2	In New York, 13 January 2008

Man-made Luxury Islands in the Gulf of Arabia

~

Green Island, Kuwait ≈ holiday island with an amphitheatre and camel racing (opened 1988)

The Pearl, Qatar ≈ $2.5bn island over 5 million square metres (2008)

Two Seas, Bahrain ≈ $3bn island built in the shape of a sea horse (2007)

Jumeirah, Jebel Ali, Deira Island, Dubai ≈ $12bn archipelago in the shape of palm branches (2002–2007)

The remotest islands

Bouvet (Norway) lies halfway between South Africa and Antarctica. The nearest land (*Gough Island*) is 1,641 kilometres away. This is according to *Guinness World Records*. We have not done the measurements, but there is a whole series of other possible candidates, from the *Amsterdam* and *St Paul* islands to *Crozet Islands*.

Maelstroms

There are certain places at sea where powerful whirlpools spin at great speed:

Saltstraumen, east of Bodø, Norway . . . . 40 km/h

Moskenesstraumen, Lofoten Islands, Norway . . . . 28 km/h
    (the 'original' maelstrom)

Old Sow, between New Brunswick and Maine, N America
    . . . . 27 km/h

Corryvreckan, Jura, Scotland . . . . 22 km/h
    (almost cost George Orwell his life)

Naruto, between Naruto and Awaji, Japan . . . . 20 km/h

Garofolo, Strait of Messina, Italy . . . . 18 km/h
    (known in mythology as *Charybdis*)

Hollywood stars who drowned

Natalie Wood (*Rebel without a Cause*), 1981, Channel Islands (US)

John Bowers (*Desire*), 1936, Channel Islands (US)

Lola Montez (*Siren of Atlantis*), 1951, in the bath

Flying Dutchmen

~

≈ Nobody was allowed to disturb Captain van Falkenburg in his cabin, for he was playing dice with the Devil. After a lucky throw by the Devil, van Falkenburg was condemned to travel the North Sea for ever – until he is redeemed by a virgin.

≈ Bernhard Focke was terrified of sailing to India, and swore that he would only put to sea with the help of the Devil. When the captain returned home laden with a rich cargo, the Devil demanded his due. But Bernhard Focke refused to pay up – and is condemned to sail the seas for ever, right down to the present.

≈ In 1611 Hendrik van der Decken, sailing home to Holland, was battling against the storms around Cape Horn. His ship was about to break up when the captain swore that he would sail on until he reached the end of the world. In another version an angel appeared to the cursing van der Decken, but the captain refused to show respect by removing his cap, and shot at the angel – whereupon the unbeliever was condemned to tread the decks of his ship for ever. In the most gruesome version, van der Decken also made a pact with the Devil: every time he came back from his travels, he would find a young girl in his house by the dyke. Before he set sail again, he would break the girl's neck and throw her into the sea. One day the Devil abducted a particularly pretty girl for the seaman from outside the church. She fought back against van der Decken; she prayed and begged for mercy. A halo appeared around her head – but this did not deter the captain from strangling her and throwing her into the sea. Even to the Devil, this seemed to be going too far, and from then on he made the head of the murderous captain's victim emerge from the waves wherever he sailed. 'Follow me, follow me,' calls the dead woman seductively to her tormentor, who can never escape her cries.

Sightings of the Flying Dutchman
~

11.7.1888 – The British warship HMS *Bacchante* encountered the restless sailor. Witnesses included the later George V. The ship's log records meeting a ship glowing red, which melted away into thin air as they approached. The rating who first spotted the ghost ship fell from the mast soon after and died.

24.1.1923 – Four seamen – Fourth Officer N. K. Stone, Second Officer Bennett, a cadet and a bosun – saw a schooner without sails in a glowing mist.

1939 – More than a hundred people witnessed the Flying Dutchman crossing False Bay in South Africa.

3.8.1942 – 9 a.m. Second Officer Davis and Third Officer Nicholas Monsarrat, a successful novelist and author of *The Cruel Sea*, crew members of the warship HMS *Jubilee*, sighted the Dutchman. In the same year, perhaps on the same occasion, the accursed sailor was seen for the last time to date, from a German submarine.

The price of salt

~

Sea salt prices per 100 g

Fleur de sel ≈ Guérande, Brittany. The classic, two euros.

Flor de sal ≈ Algarve, Portugal, from 40 cents in a large refill pack.

Alaea ≈ originally used in purification ceremonies, this reddish salt from Hawaii is now sprinkled on the barbecue. Four euros.

Ittica d'or ≈ Sicily (for bruschetta, light salads and grills). From 40 cents.

Kala Namak ≈ India (pink in colour with a faint sulphurous taste). Three euros.

Murray River ≈ Australia (a mild, peach-coloured salt which dissolves readily). 7.50 euros.

Halen Mon ≈ Wales (contains zinc, magnesium and calcium). Five euros.

Salish ≈ from NW USA (smoked on alderwood, a favourite with salmon and on barbecues). 31 cents.

Sea state (Petersen scale)

~

Wind force	Sea state
0	0 sea like a mirror
1	1 ripples
2	2 crests of glassy appearance
3	2 crests begin to break ('sea horses')
4	3 small waves
5	4 fairly frequent horses
6	5 spray
7	6 sea heaps up
8/9	7 high waves
10	8 very high waves
11	9 exceptionally high waves
12	12 heavy seas

Sea state scale of the
World Meteorological Organization

Sea state	Description	Wave height
0	calm (glassy)	0 m
1	calm (ripples)	0–0.1 m
2	smooth (wavelets)	0.1–0.5m
3	slight	0.5–1.25m
4	moderate	1.5–2.5m
5	rough	2.5–4m
6	very rough	4–6m
7	high	6–9m
8	very high	9–14m
9	phenomenal	over 14m

Wind and wave in the North Atlantic

The height, length and speed of waves vary greatly according to the sea.

Wind Strength (Beaufort)	Wind Speed (metres per second)	Wind Speed (Knots)	Sea	Height of Waves (in cm)	Length of Waves (in m)	Speed of Waves (metres per second)
0	up to 0.2	>1	0			
1	up to 1.5	1–3	1		up to 10	
2	up to 3.3	4–5	2	5–13	up to 12.5	6
3	up to 5.4	7–10	2	26–43	up to 12.5	
4	up to 7.9	11–15	3	66–93	up to 37.5	8.5
5	up to 10.7	16–21	4	127–165	up to 60	10
6	up to 13.8	22–27	5	209–316	up to 105	13
7	up to 17.1	28–33	6	444–594	up to 105	16
8	up to 20.7	34–40	7	778–977	220	19
9	up to 24.4	41–47	8	1205–1460	250	21
10	up to 28.4	48–55	8	1720+	270	21–2
11	up to 32.6	56–63	9	2020+	600	22
12	up to 36.9	64–71	9			

The poisonous weever fish

The Lesser Weever fish (*Echiichthys vipera*), which causes an excruciating wound from poisonous barbs on its back, is increasing in population around the British coast.

According to one expert, 'People can die if they go into anaphylactic shock after being stung by the weever fish so they should take precautions and not paddle barefoot.' The lesser weever, which grows to about 15cm, lives on shrimps and comes inshore to feed during the summer months. It lies buried in wet sand at low tide or in shallow water and, when disturbed, erects its black dorsal fin with venomous spines.

The pain is most intense during the first two hours, when the foot goes red and swells up and then feels numb until the following day. The pain and irritation may last for up to two weeks.

The most effective treatment is to put the affected limb in water as hot as the victim can bear, without causing scalding for at least 30 minutes. In tests, the venom, a type of protein, breaks down above 40° C. This should bring swift and permanent relief.

to cut and run – meaning 'to run away'

This phrase probably originated with an immediate risk to a ship that has no time to raise anchor. The anchor cable was cut, the anchor abandoned and the ship made sail with the utmost urgency. This definition appears in the 1794 manual *The Elements and Practice of Rigging and Seamanship* by David Steel.

copper-bottomed – meaning 'genuine, reliable, trustworthy'

This is derived from the practice of lining ships' wooden hulls with copper sheeting which prevented wood-boring insects like the teredo worm from destroying the timber. It also resisted barnacle infestation, which would reduce a ship's speed by increasing hull drag. Copper compounds are still used in anti-fouling paints for ships great and small.

hard and fast – meaning 'inflexible, without doubt, unbending'

From a ship that was firmly aground, stuck fast or beached and was not going anywhere until possibly the next high tide.

ship-shape and Bristol fashion –
meaning 'orderly and in excellent condition'

Possibly originated because the port of Bristol experienced a big tidal range of around 9 metres between high and low tides. As a result ships using the port had to be sturdily built and cargoes had to be very well stowed to avoid shifting. The phrase first appears in print in *Two Years Before the Mast*, by the American writer Richard Dana, in 1840. The famous story, based on personal experience, became a 1946 Paramount film with Alan Ladd in the lead role.

Tell it to the marines –
meaning 'a scornful reply to a fanciful story'

Marines on board ship in the 17th century were regarded by hardened old salts as naive landlubbers who would believe anything. The early version was: 'You may tell that to the marines, but sailors will not believe it.' In Anthony Trollope's *The Small House at Allington*, published in 1864, the current shorter version appears for the first time in print: 'Is that a story to tell to such a man as me! You may tell it to the marines!'

The most poisonous sea creatures

(According to the LD50, which expresses how many mg of poison per kilogramme of the victim's bodyweight are required to kill half the test population.)

Stonefish . . . . 0.2 mg per kg
Box jellyfish . . . . 0.04 mg per kg
Sea cobra . . . . 0.02 mg per kg
Blue-ringed octopus . . . . 0.02 mg per kg
Zoanthids . . . . 0.0002 mg per kg

The Wilson Cycle

In 1970 Canadian scientist John Tuzo Wilson discovered the cyclical process by which the ocean basins open and close as a result of the stretching and collision of tectonic plates.

1 *Initial phase* – There is tectonic stability. There may be a hot spot lurking beneath the continental plate, which may erupt as a volcano.
2 *Trench or rift stage* – faults, fractures, and volcanoes develop.
3 *Red Sea stage* – A trench opens, land sinks, lava flows in and forms a basalt crust all around it: a new ocean is created.
4 *Atlantic stage* – The ocean opens out and grows. Lava gushes up from the ocean floor, pushing in between the two plate boundaries

and forcing them apart. An ocean ridge is formed, a mountain range under the sea, whose highest peak (e.g. Iceland) may project above sea level.

5 *Pacific stage* – Subduction: an ocean plate slides beneath a continental plate. Where the sea crust disappears into the depths, deep sea trenches are created: the sinking rock melts – a process that sets off powerful volcanic activity.

6 *Mediterranean stage* – The ocean shrinks and shrinks, the sea bed forms folds, there is grating along the shores, mountains are formed.

7 *Himalaya stage* – The sea has completely disappeared: two continents collide. Mountains rear up; faults and fractures may produce new volcanoes.

8 *Pause in tectonic activity* – The plates have coalesced. The former boundary is called a suture zone. A new cycle can now begin.

Shrinking and expanding oceans

≈ America and Europe are drifting apart. Every year the Atlantic becomes 70 millimetres wider (about the growth rate of a fingernail).

≈ The Californian Pacific plate is pressing up against the rest of America. The Pacific expands eastwards by 60 millimetres annually.

≈ Africa and South America are pulling apart from each other at the rate of 20 – 30 millimetres a year. The distance between Africa and North America is also increasing yearly by 20 millimetres.

• The Indian plate is pushing northwards, forcing the Himalayas yet further upwards. The Indian coast is retreating northwards at 20 millimetres a year.

• The Horn of Africa is separating from the rest of the continent by ten millimetres per annum. In ten million years, at the latest, the Red Sea will flood into the Afar Depression to produce a new sea.

- Africa is pushing towards Europe, and the Mediterranean is shrinking by six millimetres year on year.

The first woman to sail around the world

~

'Saturday 28 to Sunday 29. Fine weather, quiet sea. With a good breeze from S-SSE we sailed WNW until eight in the morning, then W¼ NW for the rest of the night. Our sails were reefed, and the topsail almost completely furled, for we had to wait for the *Étoile*, which was making heavier weather of it than ever. The state of our provisions obliges us to seek out a European settlement. None the less, we are determined to press on. Our food supplies have been rationed for some time. There are still a few chickens left for the sick, and three turkeys for the next three Sundays. We had one at midday today, and all enjoyed it greatly. [There were also] these large beans, called broad beans, as well as bacon, salt and ham – but all nearly three years old!

Note. There was a strange occurrence on the *Étoile* yesterday. There has been a rumour making the rounds on both ships for some time that de Commerçon's servant, called Baré, was a woman. The rumour was fed by his slight stature; the care he took never to change his clothes when others were present and he never seemed to answer the call of nature; his voice; his beardless chin, and other signs. Suspicion turned into certainty during an incident on Cythera. Commerçon had gone ashore with Baré, who was used to assisting him with his botanizing, carrying weapons, food, reference books with such strength and stamina that he had earned the reputation of a pack mule. Hardly had they landed when they were surrounded by the islanders, shouting out that he was a woman, and offering to do him the honours of the island. The duty officer had to rescue them. By royal command, I was now charged with establishing whether the rumour was justified. Baré confessed with tears in her eyes that she was a girl, that she had already been deceiving her master by dressing as a man when she served him in Rochefort, and had previously worked as a male servant for a gentleman of Geneva. She had been born in Burgundy and orphaned at a tender age; having been ruined by the loss of a legal battle, she had decided to conceal her sex, being eager to sail around the world. She will be the first woman ever to have achieved this, and I admire her resolve all the more because she has so far committed no offence, and fulfilled all her duties to perfection. I am making sure that nothing happens to her. The court will surely forgive this infringement of regulations. Her behaviour is hardly likely to become general. She is neither pretty nor ugly, and under 25 years of age.'

(*From the journal of Louis Antoine de Bougainville, who in 1766–9 became the first Frenchman to circumnavigate the globe.*)

Classic five-master ships

Maria Rickmers, launched 1892, lost on maiden voyage.

Potosi, launched 1895, scuppered 1925 after a fire.

R. C. Rickmers, launched 1906, sent to the bottom under a British flag, 1917.

France II, launched 1911, ran aground 1922.

København, launched 1922, lost in 1928.

France I, launched 1890, sank 1901.

Only six ships have been built with five masts

These are potentially life-saving aids to memory. For instance, the interpretation of ships' lights at night is fraught with danger, so the following verse, imprinted in the watchkeeper's memory, could avoid the deadly risk of collision.

When both side lights you see ahead,
Port your helm and show your red;
But green to green and red to red,
Means perfect safety – go ahead.

> If to your starboard red appear,
> It is your duty to keep clear;
> To act as judgement says is proper,
> To port or starboard, back or stop her.

> But when upon your port is seen
> A steamer's starboard light of green,
> There's not so much for you to do,
> For green to port keeps clear of you.

'Jeanie Johnston'

The 600-ton three-masted sailing barque *Jeanie Johnston* was built by Scottish-born Quebec shipbuilder John Munn (1788–1859) in 1847. During the 1840s and 1850s the Irish famine created over 1.5 million migrants desperate to reach North America, but it could be a dangerous journey as many were transported in 'coffin ships'; in 1847, 17,465 shipboard deaths were documented. The *Jeanie Johnston* and Captain James Attridge, however, had an impeccable record, carrying 2,500 passengers on 16 voyages between 1847 and 1858 to North America without loss. On the return trips to Tralee,

County Kerry the ship carried timber and corn. The vessel, loaded with timber destined for Ireland, finally succumbed to the elements in mid-Atlantic on 31 October 1858 and sank, but all were rescued by a Dutch ship. *Jeanie Johnston*, the most famous of Irish famine ships, maintained her proud record to the last.

In 2003 an authentic replica, designed by naval architect Fred Walker, sailed from Tralee to the USA and Canada and back again after visiting 20 ports in five countries along the way. She will continue operating as a sail-training vessel with an onboard museum when in port.

Well-travelled eels

(Distances travelled by eels from the spawning ground in the Saragossa Sea to the inland waters where they will spend the next ten years until they reach sexual maturity. Then they will swim back downriver into the open sea in order to return to their birthplace.)

St Lawrence River, Canada	3820 km
Tejo, Portugal	4980 km
Shannon, Ireland	4965 km
Thjorsa, Iceland	5150 km
Loire, France	5600 km
Ijssel, Holland	7025 km
Po, Italy	8200 km
Lake Vidgan, Sweden	8300 km

Annual worldwide fish catch

(Source: *FAO, the Food and Agriculture Organization of the UNO*)

	1999	2000	2001	2002	2003
Anchovies	8,723,265	11,276,357	7,213,077	9,702,614	6,202,447
Alaskan pollock	3,277,286	2,938,230	3,144,465	2,654,854	2,887,962
Herring	2,411,408	2,381,011	1,952,605	1,853,936	2,088,744
Skipjack tuna	1,968,315	1,967,340	1,817,142	1,997,151	2,110,681

Fish you can eat with a clear conscience

(*because according to the WWF they are not endangered by over-fishing – yet*)
Wild Alaskan salmon, herring, mackerel, pollock
(but Alaskan pollock is said to be under threat)
≈ ecologically farmed salmon and shrimp ≈ freshwater fish such as zander, carp and trout

Fish you shouldn't eat

Nile perch (consumption causing species extinction) ≈ farmed salmon (fed with by-catch) ≈ shrimps, prawns, king prawns (farming destroys mangrove forests) ≈ tuna, shark, cod, flounder, red sea bass ≈ halibut ≈ hake ≈ plaice ≈ swordfish ≈ skipjack tuna ≈ eel ≈ spiky dogfish ≈ haddock (over-fished, perhaps even to near extinction).

Travellers' Century Club 'country points'

To qualify for membership of the Travellers' Century Club, you have to collect enough 'country points' by visiting at least one hundred countries. However, the club recognizes altogether 315 distinct 'countries', 123 more than the United Nations Organization has member states. The following places count separately as far as the club is concerned:

In Europe – England, Scotland, Wales, Gibraltar, Guernsey, Northern Ireland, Isle of Man, Jersey, the Canaries, the Balearics, the Azores, Madeira, Corsica, Crete, Rhodes, Sardinia, Sicily, the Ionian Islands, Istanbul, Northern Cyprus, Kaliningrad, Kosovo, Lampedusa, Montenegro, Spitzbergen, Srpska, Transdniestr, Greenland, the Faroes, the Aland Islands.

Plus the last outposts of the British Empire – Falkland Islands & British Antarctic Territory, Pitcairn, St Helena, Tristan da Cunha, British Virgin Islands, Turks & Caicos, Anguilla, Ascension, Montserrat, British Indian Ocean Territory, Cayman Islands, the Bermudas.

Plus the following French Departments and territories – French Polynesia, the Marquesas, Wallis & Futuna, French Southern and Antarctic Territories, Martinique, Guadeloupe, Réunion, Nouvelle-Calédonie, Mayotte, French Guyana, French Antilles, Europa Island, St Pierre & Miquelon

United States territories – Alaska, Hawaii, American Samoa, Midway, US Virgin Islands, Johnston Atoll, Guam, Puerto Rico, Wake Island, the Mariana Islands

In China – Hong Kong, Macao, Taiwan, Tibet, Hainan

In India – Kashmir, Sikkim, Lakshadweep, the Andaman & Nicobar Islands

In the Indonesian Archipelago – the Moluccas, the Sunda Islands, Sulawesi, Kalimantan, Sumatra, Java, Irian Jaya

Australian possessions – the Lord Howe Islands, Christmas Islands, Norfolk Island, Tasmania, the Australian Antarctic

Belonging to New Zealand – the Chatham Islands, Niue, Tokelau.

The United Arab Emirates – Abu Dhabi, Ajman, Dubai, Fujeirah, Ras Al Khaimah, Sharjah, Umm Al Qaiwain.

Canadian – Prince Edward Island

Malaysian – Sabah, Sarawak

In Africa – Spanish Morocco, Somaliland, Western Sahara, Zanzibar

In the Pacific – the Easter Islands, Galapagos, Cocos Islands, Juan Fernandez Island, Ogasawara, Okinawa, the Bismarck Archipelago, the Phoenix Islands

In the Caribbean – Aruba, Netherlands Antilles, San Andrés, Providencia.

Belonging to Mauritius – the Rodriguez Islands

The Seychelles – Zil Elwannyen Sesel (Outer Islands)

Belonging to Azerbaijan – Nakhichevan

Russian Siberia

The Antarctic sectors – Argentinian, Chilean, New Zealand, Norwegian, plus Bouvet Island.

Nautical miles and knots

Sea mile . . . . 1,852.01 metres
(exactly one 21,600th of the Equator – one minute of arc)

Knot . . . . 1 sea mile per hour
(From at least 1500, Dutch sailors were trailing overboard a piece of wood attached to a rope marked out by knots at regular intervals. The seamen counted the knots as the line ran out from a coil. After 30 seconds by the hourglass, the number of knots counted corresponded to the sea miles per hour.)

Fathom . . . . 182.88 centimetres
(The Old English word 'faethem' means 'stretched out'. Fathoms

were longer in Greece and Austria: 185.42 and 189.64 centimetres respectively.)

Cable (British) .... 185.3 metres

Cable (American) .... 219.50 metres
   (This was the term for a long rope.)

Whale and dolphin chart-toppers

~

The six best recordings of their songs:

Walstimmen. Gesänge und Rufe aus der Tiefe
   (*Whale Voices: Songs and Calls from the Deep*) (Edition Ample)
   Celebration of the Hawaiian Spinner Dolphin
   (Magical Island Sounds)
   Council of the Humpback Whales (Magical Island Sounds)
   Songs of the Humpback (Living Music)
   Dolphin Song (Animal World)
   Humpback Whales (Sounds of the Earth)

Black Star Line

~

Marcus Garvey and the back-to-Africa movement

Marcus Garvey, born in Jamaica in 1887, was the most successful leader of black people in the early decades of the 20th century. His 'Universal Negro Improvement Association and African Communities' (Imperial) League' – UNIA-ACL for short – had three million followers. His great dream was to build a sailing fleet to take his people back to Africa.

By September 1919 he had taken the first step: the red, black and green flag of the UNIA-ACL flew from the mast of the first ship of his 'Black Star Line', the *Frederick Douglass*. The captain and crew were all black, as were the thousands of shareholders who had invested in the shipping company. But even on its maiden voyage under new management, the antiquated, over-priced coal barge ran into problems with documentation, over failure to pay for an insurance policy. And then, on its first proper voyage, the captain ran the ship aground on a sandbank. The crew were about to jump ship, when the *Frederick Douglass* lifted off spontaneously. At last the five-dollar share holders of Cuba and Panama could see with their own eyes that the Black Star Line really did exist. They rushed to buy more shares.

The *Frederick Douglass* accepted a lucrative deal to export whiskey to Cuba, with the aim of beating the start of prohibition; within a few days, every vat found in US waters would be confiscated. But the *Frederick Douglass* was not built for speed. The engines were in need of reconditioning, and it did not help when the wrong valve was opened and seawater flooded in. The crew panicked and threw the vats overboard, where, fortunately, they were picked up by smaller support boats. The stricken ship was repaired, and the whiskey exporter promised an extra $2,000 if the crew took better care of its valuable consignment on the next trip. But this time the men broke into their liquid cargo, and staggered ashore in Havana dead drunk, with empty barrels. On the return journey the captain rammed his vessel into a reef, and was replaced by a white man. The shipping company, deep in debt, had to sell the *Frederick Douglass* for a mere $1,625, only a tenth of what it had cost to buy barely two years earlier. The Black Star Line's next ship was not to enjoy a smoother run, either: once again financed by the small investments of black shareholders – farmers, cleaning ladies and workmen – the *Antonio Maceo* developed engine trouble on its first two trips, and on the third a drunken engineer let seawater into the steam turbine. After astronomical repair costs of $135,000 had been run up (the ship had cost half that much to buy, and was clearly a bad bargain even at that), the *Antonio Maceo* was left to rust in a Cuban port.

In three years the Black Star Line had thus dissipated $600,000. Half of that sum had gone on provisioning, advertising for new shareholders, expenses and bribes, if it had not been embezzled. All the same, Marcus Garvey was not yet ready to abandon his dream of acquiring a ship.

However, trouble was also brewing on the other side of the Atlantic. The advance team sent to Liberia by UNIA-ACL had made itself deeply unpopular when its unflattering memorandum about the Liberian government was made public. The descendants of freed slaves from the USA were keeping the native population in a condition of serfdom, an unresolved conflict that was to unleash a bloody civil war at the end of the 20th century. Not entirely without justification, the Liberian government feared that Garvey's UNIA-ACL might be planning to seize power. So it sold out voluntarily instead to the American Firestone Rubber Company, which then ran the West African country for decades as a kind of private colony.

Marcus Garvey's vision had fallen apart; he was tried and convicted for fraud. The founder of the Black Star Line died in poverty in London in 1940. Streets are still named after him in Kingston Jamaica, and among Rastafarians he is regarded as an apostle.

The biggest animal sound archives

From sea robins* to albatrosses:
British Library, London . . . . 150,000 animal sounds
(www.bl.uk/listentonature)
Cornell Lab of Ornithology, USA . . . . 150,000 bird calls
(www.birds.cornell.edu/brp/research/index.html)
Humboldt-Universität, Berlin . . . . .110,000 sounds
(www.biologie.hu-berlin.de/~tsarchiv/index_ger.html)

*a fish that makes a croaking noise similar to a frog

Naming hurricanes

Hurricanes were first named in the West Indies after the saint's day on which the hurricane occurred. Women's names were used in the nineteenth century by an Australian meteorologist for tropical storms. The US National Weather Service began using women's names in 1953 and has only included men's names from 1979.

For Atlantic Ocean hurricanes, the names may be English, Spanish or French. The World Meteorological Organization uses six lists of 21 alternating male and female names (in alphabetical order but minus the letters Q, U, X, Y, Z) in a rotating six-year cycle. A name is removed if a hurricane has been particularly destructive and a new name is substituted.

Hurricane names for Atlantic storms 2007

Andrea	Humberto	Olga
Barry	Ingrid	Pablo
Chantal	Jerry	Rebekah
Dean	Karen	Sebastien
Erin	Lorenzo	Tanya
Felix	Melissa	Van
Gabrielle	Noel	Wendy

Tsunamis

26.1.1700 ≈ Vancouver Island, Canada, no victims
1.11.1755 ≈ Lisbon, Portugal, 100 000 victims
26.8.1883 ≈ Krakatao, Indonesia, 37 000 victims
18.11.1929 ≈ Newfoundland, Canada, 28 victims
1.4.1946 ≈ Aleutian Islands, North Pacific, 165 victims
22.5.1960 ≈ Valdivia, Chile, 3000 victims

27.3.1964 ≈ Alaska to California, 11 victims
(and a further 121 from an earthquake)
12.12.1979 ≈ Columbia and Ecuador, 259 victims
12.7.1993 ≈ Hokkaido, Japan, 300 victims
26.12.2004 ≈ Indonesia, Sri Lanka, Thailand, Bangladesh,
Malaysia, India, Maldives, 275 000 victims
17.7.2006 ≈ Java, Indonesia, 525 victims

Florida threatened by Tsunami from Canary Isle

The island of La Palma is not only the steepest island in the world but has also been the most volcanically active of the Canary Isles in the past 500 years. (The last eruption was in 1971.) A 1949 eruption caused several cubic kilometres of rock to slide a few metres towards the sea. This has also opened a two-kilometre-long fracture visible to this day.

There is a real danger that the side of the volcano facing west may collapse into the Atlantic sending a catastrophic tsunami to devastate Florida. A member of a recent expedition stated, 'The island is very unstable and this is something which could happen fairly soon.'

Greenwich – home of 0 degrees longitude

John Harrison (1693–1776), a self-taught Yorkshire clockmaker, solved the longitude problem by inventing a timepiece, unaffected by temperature, humidity and motion, that could tell the correct time at sea. His H4 chronometer, built in 1759 after years of experimentation, eventually won for John Harrison the Longitude prize from the British government.

A prime meridian was needed from where all longitude could be calculated. From 1767 sailors all over the world who relied on the nautical almanacs of Nevil Maskelyne (fifth Astronomer Royal) began to calculate their longitude from Greenwich. In Washington DC, in 1884, the representatives of 26 countries voted to make the common practice official.

However, the French were reluctant to conform until 1911 and even then could not bring themselves to use the phrase Greenwich Mean Time. Instead they used the term 'Paris Mean Time, retarded by nine minutes twenty-one seconds'.

The Freedom of the seas

Three sea miles out from shore, the range of a cannon, used to be the limit of a country's sovereignty over its seas. Today the law is that:
A state's territorial waters extend to *twelve sea miles*.
Over the first twenty-four miles *out from its coast, a state can* exercise control over customs, entry and hygiene in respect of ships entering its waters.
A state has the right to economic exploitation of the sea up to 200 *miles* from its coast. This is what gives rise to most conflicts.
59% of the sea's area is ex-territorial. There is no control over the laying of cables and pipes, fishing, scientific research, sea and air transport, or the building of artificial islands. Founding states outside territorial waters is, unfortunately, no longer possible. The nearest country can now intervene. The seabed counts as the common heritage of all humanity. The International Seabed Authority is tasked with ensuring that countries with no involvement in deep-sea mining nonetheless have a share in the returns.

Disputed islands

Who lays claim to what?
The list is long. There's clearly some tidying up to be done …

Spratly ≈ China | Vietnam | Malaysia | Philippines | Taiwan | possibly also Brunei

Bolshoi ≈ Russia | China

Etorofu, Kunashiri, Shikotan, Habomai ≈ Japan | Russia

Senkaku/Diaoyuta ≈ China | Taiwan | Japan

Pratas, Paracel, Macclesfield Bank ≈ China | Taiwan | Vietnam

Takeshima/Tokdo ≈ Japan | South Korea

Nipa ≈ Indonesia | Singapore | Vietnam

Koh Pich ≈ Cambodia | Vietnam

Marore, Miangas, Marampit ≈ Indonesia | Philippines

Berhala ≈ Indonesia | Malaysia

Rondo ≈ Indonesia | India

South Talpatty/New Moore ≈ Bangladesh | India

Batek, Dana ≈ East Timor | Indonesia | Australia

Île Matthew, Île Hunter ≈ France | Vanuatu | New Caledonia

Chagos Archipel ≈ Seychelles | Mauritius | British Indian Ocean Territory

Diego Garcia ≈ Mauritius | British Indian Ocean Territory

Bassas da India, Europa Island, Glorioso, Juan de Nova ≈ Madagascar | France

Tromelin Island ≈ Madagascar | Mauritius | France

Mayotte ≈ France | Comores

Wake Island ≈ Marshall Islands | USA

Lam, Khan, Khinok ≈ Thailand | Burma

Abu Musa, Lesser and Greater Tumb Islands ≈ Iran | United Arab Emirates

Hawar ≈ Bahrain | Qatar

Qaruh, Umm al Maradim ≈ Kuwait | Saudi Arabia

Zuqar-Hanish Archipelago ≈ Yemen | Eritrea
Peñón de Alhucemas, Peñón de Vélez de la Gomera, Islas Chafarinas
  ≈ Spain | Morocco
Mbanie, Coctotiers, Congas ≈ Gabon | Equitorial Guinea
Kasikili, Situngu ≈ Namibia | South Africa
San Andrés Archipel, Quita Sueno Bank ≈ Columbia | Nicaragua
Navassa Island ≈ Haiti | USA
Machias Seal Island ≈ USA | Canada
Rockall ≈ Great Britain | Ireland | Denmark | Iceland
Falkland Islands/Malvinas, South Georgia, South Sandwich Islands
  ≈ Great Britain | Argentina
Kardak/Imia ≈ Greece | Turkey
North Cyprus ≈ Turkey | Cyprus

(*In addition to all these, Niger, Uganda, Congo, and Brazil are
engaged in disputes over islands in rivers; nor is there agreement over
borders in the Antarctic. And how the international border between
Switzerland, Austria and Germany divides Lake Constance has never
been settled.*)

What is an island?

~

According
to the UN Maritime Law
Conference, an island is defined as a
land area of natural origin surrounded by
water and projecting above sea level at high tide.
Japan has spent over £330m on fortifying the
Okinotorishima islets in order to assert its right to a
400,000 square metre Economic Exclusion Zone. (China
claims that they are not islands, but rocks, which
cannot be the basis for proclaiming an EEZ.) New
islands can be created through the building up
of icebergs, through volcanic eruptions,
or through the disintegration of
old islands.

New islands on the map

~

Tuluman, Papua-New Guinea 1960
Surtsey, Iceland 1963
South Talpatty/New Moore, Bangladesh/India 1974

The ship that was saved by a poem

~

The USS *Constitution*, or 'Old Ironsides', a 2,200-ton three-masted
wooden frigate launched in 1797, is the oldest commissioned ship
afloat in the world and is still a US navy vessel. (HMS *Victory*
(Portsmouth, England)is the oldest commissioned ship still extant,
but she is permanently berthed in a dry dock.)

In 1830, after her long and distinguished active service, the US navy proposed to break up and scrap the old ship, but an emotional poem, 'Old Ironsides' by 21-year-old Oliver Wendell Holmes (1809–85) printed in the *Boston Daily Advertiser* in September 1830, stirred the nation to outrage, thus saving her for posterity.

For who could fail to be moved by lines like these:

> Her deck, once red with heroes' blood ...
> No more shall feel the victor's tread
> Or know the conquered knee;
> The harpies of the shore shall pluck
> The eagle of the sea!

The 'meteor of the ocean air' can be found at Pier 1 at Charlestown Navy Yard, Boston. On her 200th anniversary the *Constitution* set sail for the first time in over a hundred years and still occasionally puts out to sea to 'sweep the clouds' once more.

Words for snow among the Inupiat Inuit of Alaska

~

ainu, apun: snow

aniu: packed snow

mauyaq: to battle through snow

nutabaq: powdery snow

qiqsruqaq: glassy snow during a thaw

sitxig: hard snow

niefiqsimaruq: to use snow to stick igloo blocks together

illuktuq: snow-blind

apiruq, apigaa: snow-covered

piqsiqsuq: snowstorm

qanniqsuk, qanigruaqtuq: snowing without wind

qatiqsubniq: light snow through which you can walk

auksalaq, auksajaq: melting snow

aukkaa: melting

qayuqjaq: ribbed surface of snow

aputyaq: igloo

natibviksuq: sliding snow

apuyyaq: area of snow

uniuvak: bank of snow

qimaugruk: snowdrift

mapsuq, mavsa: threat of melting/sliding snow

qannik: snowflake

nuluq: wickerwork of snowshoe

tagluk: snowshoe

sisuuq: avalanche

aqixuqqaq: powdery snow

mitaixaq: light snow over a frozen ice-hole

mauruq: to

pukak: snow like sugar

puuvruktuq: to tramp through snow

mixik: whirling snow

Buried at sea

Maria Callas ≈ John F. Kennedy Jr ≈ L. Ron Hubbard ≈ Francis Drake ≈ Philippe Cousteau ≈ Ingrid Bergmann ≈ Rock Hudson Robert Mitchum ≈ Steve McQueen ≈ Vincent Price Janis Joplin ≈ Dennis Wilson ≈ Jerry Garcia

Billy, Ben and Herman

Benjamin Britten (1913–1976) composed his opera *Billy Budd* in 1951 and it was first performed at the Aldeburgh Festival as part of that year's 'Festival of Britain' celebrations. The opera is based on a novel begun about 1886 by Herman Melville (1819–91) but only discovered in manuscript form among the author's papers in 1924.

The story is set in 1797 on board the ship HMS *Bellipotent* at the time of naval mutinies and Napoleon's military ambitions. Billy, falsely accused of incitement to mutiny but in fact an admirable character loved by the crew, is court-martialled for unintentionally killing depraved John Claggart, the ship's master-at-arms. Captain Vere, despite Billy's obvious innocence, persuades the court-martial panel to give a guilty verdict and Billy is doomed to be hanged from the ship's yardarm.

In 1960 Britten revised the work, reducing it from four acts to two, and it is this revised form that it is now generally performed. The librettists, E. M. Forster (1879–1970) and Eric Crozier (1914–94), transformed the story into an opera of great dramatic power and psychological subtlety. The focus of the opera is the moral dilemma facing Captain Vere, who has to decide between saving Billy or following what he sees as his duty as ship's captain to uphold discipline.

Grace Darling, heroine of the sea

Grace Horsley Darling, born in Bamburgh, Northumberland, in 1815, was the daughter of William Darling, the keeper of the Longstone Light on the Farne Islands. On 7 September 1838 the newly built steamer *Forfarshire*, on passage from Hull to Dundee with 63 passengers and crew on board, foundered on the rocks and started breaking up. William Wordsworth takes up the story:

> With quick glance
> Daughter and Sire through optic-glass discern,
> Clinging about the remnant of this Ship,
> Creatures – how precious in the Maiden's sight!

Against all the odds, nine survivors were rescued by father and daughter, who became household names. No less a figure than William Topaz McGonagall (1822–1902), widely regarded as the world's best bad poet, was inspired to celebrate this national treasure's achievement:

> And as the little boat to the sufferers drew near,
> Poor souls, they tried to raise a cheer;
> But as they gazed upon the heroic Grace,
> The big tears trickled down each sufferer's face.

Grace Darling died in 1842. Her ornate and imposing tomb can be found in St Aidan's churchyard, Bamburgh.

Where oceans meet

Atlantic/Indian Ocean – from Cape Agulhas (South Africa) along the line of longitude 20 E to Antarctica.

Atlantic/Pacific – from Cape Horn along 67° 16′ W to Graham Land, Antarctica.

Pacific/Indian Ocean – from the Strait of Malacca (Indonesia/ Malaysia), then between Sumatra and Java, continuing between the Little Sunda Islands and Timor: then across the Timor Straits to Cape Londonderry (or Cape Talbot) in Australia, and from South-East Point (Australia) along 146° 55′ E to Antarctica.

The Arctic Ocean – from Nordkapp (Norway) to Øyrlandsoden (Spitzbergen), on to Borgarfjordur (Iceland), and then from Hornbjarg (Iceland) across the Denmark Strait to Cape Hammer in Greenland; continuing via Disko Island on the western side of Greenland along latitude 74 to Baffin Land in Canada; then south along the coast via Resolution Island to Button Island – thereby including Hudson Bay – and then across the narrowest point of the Bering Strait and along the Siberian coast back to Nordkapp.

Southern Ocean – This is no longer regarded as a separate ocean. Its borders were very straightforward, however: everything south of latitude 60.

Ocean depths

Pacific ≈ 11,034m, in 1957 the Soviet research vessel *Vityaz* measured a depth of 11,034 metres, known as the *Mariana Hollow*. It is in the Mariana Trench, 1850 km east of the Philippines, and is the deepest point in the world's oceans.

Atlantic ≈ 9219m, Milwaukee Deep, 150km north of Puerto Rico, named after the USS *Milwaukee*.

Indian Ocean ≈ 8047 m, Diamatina Deep, 1125 km SW of Perth, Australia, named after the Australian Navy research vessel.

Arctic Sea ≈ Litke Deep, 5449 m, 220 km SW of Nordaustlandet (Norway).

Mediterranean ≈ 5267 m, Calypso Deep, S of the Peloponnese (Greece).

Black Sea ≈ 2244 m.

North Sea ≈ 725 m in the Norwegian Trench.

Baltic Sea ≈ 459 m, Landsort Deep, between Södertörn and Gotland.

'Bruce' the mechanical shark

'Bruce' was the nickname given by the *Jaws* film crew to the giant mechanical shark which had to stand in for the real thing. In fact there were three Bruces: one with machinery exposed on the left side; one with machinery exposed on the right; and a complete shark for overhead shots. The first two sharks were full of hydraulic, pneumatic and electronic equipment and were attached by an articulated boom to an underwater platform.

The shark's movements were remote-controlled but, with over 1500 m of plastic tubing, 25 remote controlled valves and 20 electric and pneumatic hoses powering various moving parts of the beast, there was plenty of scope for malfunctions. Seawater corroded its metal parts, requiring an almost endless series of repairs, while the 'skin' had to be replaced every week because of the bleaching effect of the sun. Each 'Bruce' had to have two sets of teeth: one made of metal for the timber-munching scenes and the other of rubber, which were more suitable for stuntmen-munching. Each shark cost $250,000 to build and over $1 million (1975 prices) to operate during the filming.

The geniuses behind all this were Joe Alves, the film's production designer, and Bob Mattey, who had built another famous sea monster, the giant squid in *20,000 Leagues Under the Sea*.

The Johnny Weissmuller factor

Swimmers who won Olympic gold and went on to have a Hollywood career:

- ≈ 8 gold medals, Matt Biondi, 1984–92 – *Dolphins, Whales & Us*
- ≈ 7 gold medals, Mark Spitz, 1972 – *Challenge of a Lifetime* (TV), *Emergency* (a TV series: Mark Spitz and his wife appear in the 8th episode, *Quicker than the Eye*)
- ≈ 6 gold medals, Michael Phelps – *Miss USA 2005*.
- ≈ 5 gold medals, Johnny Weissmuller, 1924–28 – twelve *Tarzan* films, thirteen appearances as Jungle Jim in the TV series of that name, and six other films.
- ≈ 3 gold medals, Duke Kahanamoku, 1912–20 – twelve films.
- ≈ 1 gold medal, Buster Crabbe, 1932 – 112 films (from *Tarzan* to *Alien Dead*)

Footnote: Esther Williams, for whom a film genre – the water musical – was specially created, was a swimming champion when she was a teenager: she took part in the 1936 Olympics and qualified for the 1940 games, which never took place.

Note that the Ian Thorpe who played as Erroll Flynn's double alongside Jack Nicholson in *The Two Jakes* is the namesake of the Australian five-time Olympic gold medallist.

Men's Olympic 100 m freestyle champions

Year	Time	Champion
1896	1:22.2	Alfred Hajos (Hungary)
1908	1:05.6	Charles Daniels (USA)
1912	1:03.4	Duke Kahanamoku (USA)
1920	1:00.4	Duke Kahanamoku (USA)
1924	0:59.0	Johnny Weissmuller (USA)
1928	0:58.6	Johnny Weissmuller (USA)
1932	0:58.2	Yasuji Miyazaki (Japan)

1936	0:57.6	Ferenc Csik (Hungary)
1948	0:57.3	Walter Ris (USA)
1952	0:57,4	Clarke Scholes (USA)
1956	0:55.4	John Hinricks (Australia)
1960	0:55.2	John Devitt (Australia)
1964	0:53.4	Don Schollander (USA)
1968	0:52.2	Michael Wenden (Australia)
1972	0:51.22	Mark Spitz (USA)
1976	0:49.99	Jim Montgomery (USA)
1980	0:50.40	Jörg Woithe (GDR)
1984	0:49.80	Ambrose Gaines (USA)
1988	0:49.62	Matt Biondi (USA)
1992	0:49.02	Aleksandr Popov (Russia)
1996	0:48.74	Aleksandr Popov (Russia)
2000	0:48.30	Pieter van der Hoogenband (Netherlands)
2004	0:48.17	Pieter van der Hoogenband

(Netherlands); i.e. 58.6% of the time taken by the 1896 gold medallist

Women's Olympic 100 m freestyle champions

1912	1:22.2	Fanny Durack (Australia);

22.87% slower than the men's gold medallist in the same year

1920	1:13.6	Ethelda Bleibtrey (USA)
1924	1:12.4	Ethel Lackie (USA)
1928	1:11.0	Albina Osipowich (USA)
1932	1:06.8	Helene Madison (USA)
1936	1:05.9	Rie Mastenbroek (Netherlands)
1948	1:06.3	Greta Andersen (Denmark)
1952	1:06.3	Katalin Szöke (Hungary)
1956	1:02.0	Dawn Fraser (Australia)
1960	1:01.2	Dawn Fraser (Australia)

1964 . . . . 0:59.5 . . . . .Dawn Fraser (Australia)
1968 . . . . .1:00.0 . . . . .Jan Henne (USA)
1972. . . . . 0:58.59 . . . . .Sandra Neilson (USA)
1976. . . . . 0:55.65 . . . . .Kornelia Ender (GDR)
1980 . . . . 0:54.79 . . . . .Barbara Krause (GDR)
1984. . . . . 0:55.92 . . . . .Nancy Hogshead (USA)
1988. . . . . 0:54.93 . . . . .Kristin Otto (GDR)
1992. . . . . 0:54.64 . . . . Yong Zhuang (China)
1996 . . . . 0:54.50 . . . . Jingyi Le (China)
2000 . . . . 0:53.83 . . . . Inge de Bruijn (Netherlands)
2004 . . . . 0:53.84 . . . . Jodie Henry (Australia);
or 34.5% faster than the 1912 champion and 10.5% slower than
the 2004 men's champion

Pioneering submariners

1578 – *William Bourne* was the first to have the idea of a boat that
could travel under water. Leather bags in the bilge were supposed
to take on water for submerging, while screw presses would force
it out again for surfacing. Apparently this original submarine was
supposed to be powered by oarsmen, but the air pipe, which was
meant to protrude above the surface like a drinking straw, could
never have supplied them with enough air. Fortunately, the idea
never left the drawing board.

1624 – *Cornelius van Drebbel*'s submarine actually seems to have
existed. It is even said that King James I of England was given a trial
run in it. The unfortunate rowers had to make the boat dive by
sheer muscle power.

1696 – *Denis Papin*, a French mathematics professor and inventor
of the pressure cooker, commissioned two submarines, only the
first of which was ever built. It looked rather like a watering can:

air and water pumps were supposed to control submerging and surfacing.

1773 – A carpenter by the name of *J. Day* prompted the first rescue action by the British Royal Navy following a diving accident: during the second trial of his submarine, the vessel broke up at a depth of 40 metres, under the weight of the large rocks Day was using as ballast.

1776 – During the American War of Independence, *David Bushnell*'s one-man-submarine snorkelled up the Hudson River with the aim of blowing up HMS *Eagle*, the British flagship. Probably because of lack of oxygen, Sergeant Ezra Lee in his hand-cranked, oak-and-iron construction never got close enough to attach his bomb to the enemy ship. Two other attempted under-water attacks also failed.

1800 – *Robert Fulton* was a pacifist who wanted to eliminate the world's navies with his submarines. His *Nautilus* consisted of an iron frame clad in copper sheeting. On the surface the ship was propelled by sail, and underwater rowers had to pull on straps. The British and Dutch thought the price of the craft was too high, so the inventor simultaneously opened negotiations with the other side, i.e. Napoleon. During a demonstration for the British Admiralty he blew a brig out of the water. The conservative old sea dogs were appalled by this cowardly style of warfare, and further-more believed that the *Nautilus* was a greater danger to its own side than to any enemy.

1850 – The credit for the first successful deployment of a subma-rine falls to a Bavarian, the artillery sergeant *Wilhelm Bauer*. His *Brandtaucher* (incendiary diver) was a tin can propelled by two men turning large treadwheels attached to a propeller: it was manoeuvred with the use of weights. Bauer managed to break through the Danish blockade of the Prussian coast. But during a dive in the harbour at Kiel, the thin panels at the stern broke, and Bauer ordered the complete flooding of the submarine. As the pressure was equalized, the blocked hatches could easily be opened, and the crew escaped. Thus they were the first to carry out an emergency evacuation of a sunk submarine. Bauer designed another submarine, for the British navy, but it was never built. However, he did construct the *Seeteufel* (Sea Devil) for the Russian Tsar. At the coronation of Alexander II in 1855, musicians climbed down into the 17.80-metre-long metal hull to play the national anthem. It is said to have sounded diabolical.

1863 – The Confederate submarine *David* was a kind of manned torpedo; on its very first dive it was flooded by the wake of a steamship, and sank. The next time the boat was deployed it managed to damage a ship, but the *David* – so called because it was meant to sink giants of the sea – sank with all souls. A slightly improved version, fitted with a hand-cranking device like that of the *Brandtaucher*, sank three times, drowning 23 men, including its inventor, Horace Lawson Hunley, after whom the submarine was then posthumously named: the CSS *Hunley*. On its next operation, on 17 February 1864, the *Hunley* succeeded in sinking the USS *Housatonic*. The *Hunley* has therefore gone down in history as the first submarine to succeed in sinking an enemy ship. Unfortunately, the *Hunley* itself was torn apart by the shock wave from the explosion.

1880 – The *Resurgam*, invented by the Liverpool clergyman George Garrett, was fitted with a Lamm steam engine (as used on the London Underground at the time). When the boat submerged, the engine fires were extinguished, and the submarine was driven only by the steam pressure that had previously been built up in its massive boiler. Under water, the *Resurgam* could achieve a respectable three knots. It sank on 26 February 1880 during trials off the Welsh coast.

The USS *Squalus* and Vice Admiral 'Swede' Momsen

On 23 May 1939 the submarine USS *Squalus* (SS-192), after leaving the Portsmouth Navy Dockyard, New Hampshire, sank in 73 m of water in the area known as the Isles of Shoals. All would have been lost but for the pioneering efforts of Commander Charles 'Swede' Momsen (1896–1967, inventor of both the 'Momsen Lung' and a submarine rescue diving bell later known as the McCann rescue chamber; he also developed the use of a helium/oxygen mix for safer deep-sea diving).

With 'Swede' Momsen in charge, the rescue diving chamber was lowered four times to the sub to save the 32 crew and 1 civilian, though 26 were lost in the aft compartments, which flooded as a result of a valve failure. The rescue was completed in the nick of time in spite of the lifting cable fraying to a single strand on the final dive.

In September 1939 the sub was raised and by May the following year had been restored to duty as USS *Sailfish*. After twelve eventful war patrols during World War II the sub was finally struck off charge from the navy list in April 1948 and scrapped, minus the bridge and the conning tower; they now reside at the Portsmouth Naval Shipyard as a permanent memorial.

In 2001 a TV film, Submerged, *with Sam Neill as Momsen, re-created the story of the* Squalus. (*The 'Swedes'' family actually came from Denmark.*)

The price of skins

Ray . . . . from 8 euros per skin, 37 euros for a purse
Shark . . . . 160 euros per skin, 50 euros for a purse
Nile perch . . . . 36 euros for a briefcase
Parrot fish . . . . $99 (Aus) for a handbag
Baramundi . . . . $48.50 (Aus) for a briefcase
Eel . . . . 80 euros for a briefcase
Seasnake . . . . around 80 euros for a briefcase
Salmon . . . . 20 euros per skin
Crocodile . . . . 130 euros for a briefcase
Alligator . . . . .55 euros for a briefcase

Blundering admirals

Admiral *George Tyron*, on a clear day during an exercise, ordered the ships under his command, HMS *Victoria* and HMS *Camperdown*, to execute a turning manoeuvre simultaneously and towards each other. Many of his subordinates queried the manoeuvre, since the ships were only 1100 metres apart, and each needed a turning distance of 730 metres. In the inevitable collision 358 men were drowned – including Admiral Tyron.

Vice Admiral *Pierre Charles de Villeneuve* had 19 ships of the line under his command, together with a number of frigates. The British force opposing him had only nine ships of the line and two frigates. But instead of attacking the British off Barbados, de Villeneuve fled across the Atlantic, thus ruining Napoleon's plan to invade England while the British fleet was away. The British pursued de Villeneuve and captured two ships before the battle of Cape Finisterre had to be broken off because of bad weather. Although the French fleet was now joined by another ten ships, de Villeneuve retreated (against Napoleon's orders) to Cadiz, which was promptly blockaded by the numerically far inferior British. Napoleon then demanded that his admiral break out towards the Mediterranean and Naples, so that he could at least support a landing of 4,000 soldiers there. It was only when Villeneuve learned that Napoleon was about to replace him, and that his successor was already on his way, that he put out to sea. However, when he did so he took a course that left the British plenty of time to work out a brilliant strategy. With his 27 ships, Nelson attacked the numerically stronger French fleet of 33 battleships. The battle was Trafalgar. It sealed the demise of France as a world power.

Lieutenant Commander Geoffrey Spicer-Simson, a desk-bound naval man and the oldest Lieutenant Commander in the Royal Navy, volunteered to take two gunboats overland to Lake Tanganyika. A German expedition had earlier delivered two ships there, and since then the *Hedwig* and the *Graf von Götzen* had dominated the strategically important lake. Spicer-Simson was tattooed from head to foot, and the Africans were fascinated by the rippling snakes that covered his skin. The Lieutenant Commander was obviously happy to show them off, for he believed in swimming in the nude. The display of decorative reptiles was also visible when he was dressed, because he liked to wear airy clothes; not a sarong or a kilt, but a grass skirt. He smoked cigars bearing his monogram, and he was addicted to Worcester sauce (as an aperitif). He was also notorious for making jokes that nobody understood. He called his two gunboats *Mimi* and *Toutou*, French names for a cat and a dog respectively (the Admiralty having refused to let him call the boats *Cat* and *Dog*). Having got the two boats to the lake, he took them into battle. The *Hedwig* presented no problem, but when the much larger *Gotzen* appeared on the scene, Spicer-Simson turned tail. Only later did he discover that the fearsome guns on the *Götzen* were wooden dummies (its real guns were needed by the German army).

Under the name of *Liemba*, the *Götzen* is still in service as a ferry on Lake Tanganyika. John Houston made a film, starring Katherine Hepburn and Humphrey Bogart, that was inspired by Spicer-Simson's expedition. However, the director of *The African Queen* did not draw upon the bizarre characteristics of the British commander.

Amazing voyages

Set adrift by the mutineers of the *Bounty*, *William Bligh* and his 18 loyal crew members sailed in the ship's launch, with minimal provisions and only basic navigational aids, from Pitcairn to Timor – a distance of some 6000 kilometres. Just one man died, weeks later, from exhaustion.

The *Endurance* lay for ten months trapped by the ice in Antarctic waters, until the ship was eventually crushed by the mass of ice. *Ernest Henry Shackleton* and five companions sailed to South Georgia to fetch help – 1300 kilometres in an open ship's boat.

Franz Romer, a farm lad from Lake Constance, laboriously worked his way up from cabin boy on the Elbe waterway to captain on the high seas. But his real dream was different: he gave in his notice to Hapag-Lloyd and had a sea kayak built by the firm of Klepper. On 28 March 1928 he set out from Lisbon in his 'Deutscher Sport' model. He was more than once attacked by sharks, suffered from scurvy, and overcame suicidal thoughts by self-hypnosis. After 7135 kilometres and 58 days at sea, Romer reached St Thomas in the Virgin Islands. Continuing his voyage, he ran into a hurricane near Florida, and disappeared.

Only the radio on board distinguished the balsa-wood raft *Kon-Tiki* from its predecessors of a thousand years ago, supposedly used by pre-Colombian South Americans to reach distant South Sea islands. To test this theory, Thor Heyerdahl set off with seven companions and a parrot on 28 April 1947. The adventurers drifted westwards for 101 days, and finally made landfall in French Polynesia. Today it is known that only the Easter Islands were settled from South America, not the islands of the South Pacific.

'The most successful failure in the history of exploration'

In 1893 the Norwegian explorer Fridtjof Nansen (1861-1930) set sail for the North Pole in the 800-ton *Fram*. The ship's hull was specially designed to be forced upwards out of the ice when the hull was squeezed by ice pressure. Nansen's aim was to drift across the North Pole with the sea ice. After a year in the ice (at 86° 4' N) it became clear that the *Fram* would not reach the Pole, so Nansen and Hjalmar Johansen (1867-1913) set off northwards with 3 sledges, 2 kayaks and 28 dogs on 14 March 1895.

At 86° 14' N, they had to give up when they realized that, while they were walking north, the ice they were walking on was moving south. They could not reach the *Fram* so they travelled south in a desperate attempt to find land. After 100 days and 480 km, lost and having run out of provisions and killed the last two dogs, they took to the kayaks. In July 1895 Nansen and Johansen came across a number of islands; with the five-month Arctic winter and darkness approaching, they had to build a shelter by digging a hole in the permafrost and using stones and walrus skins. They laid in a stock of walrus blubber and polar bear meat and settled in for the ultimate endurance test.

They emerged in May 1896 and headed in what they hoped was the direction of Spitzbergen. After a month they were still lost but, in one of the most extraordinary strokes of luck in the history of exploration, they heard the sound of dogs. Frederick George

Jackson (1860–1938) of the Jackson–Harmsworth Arctic expedition appeared. He told the two intrepid survivors that they had inadvertently reached Franz Josef Land (an archipelago of 191 islands). They gratefully accepted a lift back to Barentsburg, Norway where they arrived in August 1896 aboard Jackson's ship, the *Windward*.

Meanwhile, the *Fram* broke free from the ice and arrived home in the same month. To view the remains of Nansen and Johansen's hut, visit Cape Norwegia, Jackson Island, Franz Josef Land. The *Fram* can be seen at the Frammuseet, Oslo, next door to the *Kon-Tiki* museum.

Distance from land

If you want to see as much water as possible on your voyage round the world, you should keep to latitude 5 N, or longitude 170 W. Latitude 60 in the southern hemisphere is much shorter in terms of distance, but – apart from a few tiny islands – completely free of land.

The cat o' nine tails

In 1864 seamen in the British navy were subject to the following punishments: Drunkenness 24 lashes; theft 36. The maximum punishment was 48 strokes. For more serious transgressions, such as murder and sodomy, the punishment was hanging. There were significant variations in the severity of the punishment depending on where the fleet was located.

Here for example are the punishments administered to cabin boys (aged 13 to 17):

Fleet	No. of cabin boys punished	Total number of strokes	Average no. of strokes administered
Australia	12	504	42
Pacific	5	204	40.8
China	22	738	33.5
N America	42	1383	33
West coast of Africa	16	522	32.6
Cape Horn	11	354	32.2
English Channel	17	520	30.5
Mediterranean	61	1843	30.2
Coast guard	7	180	25.7
Caribbean	11	253	23

Stink balls and other missiles

~

In his classic *The Sea Gunner* (1691), John Seller gave a few practical tips for fighting sea battles:

Fire-pots, to be thrown by hand, are assembled using a clay pot filled with a pound each of gunpowder, petroleum, sulphur and sal ammoniac, together with 4 oz of camphor. Stir well and top off with hot pitch. Try lighting a small sample first; if it burns too slowly, add more gunpowder.

For *stink balls* you will need ten parts gunpowder, six parts pitch, 20 parts tar, 20 parts saltpetre and four parts sulphur. In Seller's words: 'Melt these over a soft fire together, and being well melted put 2 lbs of Coledust of the Filings of Horses Hoofs, 6 lbs Assa Fætida [a smelly resin from Persia], 3 lbs Sagapenum, 1 lb Spatula Fætida [an equally malodorous gum from Persia]. Incorporate them well together and put into this matter so prepared old Linnen or Woolen Cloath, or Hemp or Tow as

much as will drink up all this | you please, and being thrown
matter, and of these make them | between Decks will be a great
up in Balls of what bigness | annoyance to the enemy.'

PT 109 and Barbara II

Lt John F. Kennedy, USN (future 35th president) took command of an Elco Type Motor Torpedo Boat *PT 109* on 23 April 1943. She was 24.4 m long with a maximum speed of 41 knots. Armament consisted of four 21-inch torpedo tubes, two 20 mm guns, four .50-inch guns and a 37 mm gun. There was usually a crew of 3 officers and 14 men.

On the night of 1/2 August 1943, *PT 109* was on patrol at night near New Georgia in the Solomon Islands when she was cut in half by the Japanese destroyer *Amagiri*. The only hope for the 11 survivors (2 injured) was to reach an island 3 miles away. After spending 15 hours in the water they all made it to the island, JFK having towed one of the injured crew all the way despite a back injury. They were eventually rescued by *PT 157* after natives delivered a message, inscribed on a coconut, which reached Australian coastwatcher Lt Arthur Evans.

John F. Kennedy (1917–63) was awarded the Navy and Marine Corps Medal for his heroics in the rescue of the crew of *PT 109*, and the Purple Heart.

Lt George Bush, USNR (future 41st president) flew *Barbara II*, a Grumman Avenger Type TBM-1C for the last time on 2 September 1944 from the USS *San Jacinto*. The TBM Avenger 1C had a wingspan of 16.5 m and a maximum speed of 260 m/h at 4500 m. Armament consisted of two wing-mounted .50-inch guns, one dorsal turret .50-inch gun, one ventral .30-inch gun and up to 900 kg of bombs. There were three crew: pilot, rear gunner and radioman.

On 2 September 1944 four Avengers were launched from the USS *San Jacinto* to attack a radio transmitter on Chichi Island, one of the Bonin Islands about 600 miles south of Tokyo. They encountered intense anti-aircraft fire but Bush flying *Barbara II* hit the target despite damage to his aircraft. With his engine on fire he ordered the crew to bail out, but he was the only survivor to be picked up by US submarine USS *Finback* after four hours in a life raft.

George Bush Snr (1924–) flew 58 combat missions in 1944 and was awarded the Distinguished Flying Cross, three Air Medals and the Presidential Unit Citation awarded to USS *San Jacinto*.

Naval rations on British ships in the 18th century

Sunday	Monday	Tuesday	Wednesday
.45 kilograms biscuit	.45 kilograms biscuit	.45 kilograms biscuit	.45 kilograms biscuit
1.2 gallons beer	1.2 gallons beer	1.2 gallons beer	1.2 gallons beer
.52 gallons pudding	.26 gallons porridge	.9 kilograms meat	.06 gallons pudding
0.05 kilograms butter			.13 gallons porridge
.2 kilograms cheese			.05 kilograms butter
			.2 kilograms cheese

Thursday	Friday	Saturday
As for Sunday	As for Wednesday	As for Tuesday

The US Navy's first ship

The US Navy can trace its origins back to 13 October 1775, when an act of the Continental Congress authorized the first ship of a new navy for the United Colonies, as they were then known. As if to emphasize the ties that many Americans still felt to Britain the first ship of the new Continental navy was named *Alfred* in honour of Alfred the Great, King of Wessex, who is credited with building the first English naval force.

Vancouver's epic voyage

Captain George Vancouver, RN (1757–98) born in King's Lynn, Norfolk, completed one of the most brilliant marine surveying expeditions of all time during his voyages in the Pacific between 1791 and 1795. His mission was to explore the north-west American coast, negotiate with the Spanish for the return of disputed territory, chart the dangerous coastline and facilitate trade. Vancouver took his two ships, *Discovery* and *Chatham*, along the coasts of Washington and Oregon northwards to British Columbia.

In addition to Vancouver in British Columbia, about one hundred and fifty other American and Canadian place names, many reflecting Norfolk roots, were chosen by Vancouver and are still in use today. His charts are a testimony to his skill and, despite dangers and privations, he returned home having lost few men.

The 250th anniversary of his birth was celebrated in King's Lynn in 2007 with a festival culminating in the arrival of the tall ship *Earl of Pembroke*.

Chesapeake Bay

The Bay is named after the Chesapeake tribe of south-east
Virginia, believed to have been wiped out by the Powhatan tribe
a few years before the settlement of Jamestown, Virginia, by
English colonists in 1607.

Chesapeake Bay is the largest estuary in the United States.
It is about 320 km long with a maximum width of 55 km.

The surface area of the Bay and its tidal tributaries is about
11,655 square kilometres.

The Bay and its tidal tributaries have an average depth of about
6 metres with some deep troughs along most of the Bay's length
of up to 53 metres.

Of the 5 important North Atlantic United States ports, 2 are situated in the Bay – Hampton Roads and Baltimore.

One million waterfowl spend winter in the Bay's basin.

Aquatic comic heroes

Sailor Moon – a tardy heroine in Manga comics, who cries a lot and
has a black talking cat called Luna.

Spongebob Squarepants – a sponge who lives in a pineapple on the
floor of the ocean in 'Bikini Bottom', the city in the *Nickelodeon*
cartoon series.

Captain Haddock – the black-bearded, frequently drunk sailor in
the *Tintin* series. Famous for his cursing.

Popeye – the one-eyed daredevil and quick-witted philanthropist
with philosophical leanings: 'I yam whad I yam – whad am I?'

Hello, buoys!

'Argo' oceanographic profiling floats look like floating fountain pens. ('Argo' stands for 'Array for Real-time Geostrophic Oceanography', part of the global observation strategy.) They have been placed in every ocean to measure the salinity and temperature of upper waters. If you want to follow the drift of the buoys or their distribution across the seven seas, click on the Argo homepage: www.argo.ucsd.edu. For the history of 'neutrally buoyant floats', look on the web under 'Southampton Oceanography Centre'.

Some fish which live in brackish water

Wrestling halfbeak ≈ mudskipper ≈ anableps anableps ≈ (four-eyed fish) ≈ humpback puffer ≈ sailfin molly

Why is the sea salty?

Rainwater leaches all kinds of substances out of earth and rocks: potassium, calcium, silicon and aluminium, as well as sodium – of which our common or cooking salt consists, i.e. sodium chloride, or NaCl. Every year 2.75 billion tonnes of salt are washed into the sea by rivers; 50,000 trillion tonnes of salt in the sea serve to give it its familiar flavour and its average salt content of 35 parts per thousand.

Why doesn't the sea get saltier and saltier?

Thousands of cubic kilometres of seawater evaporate daily, leaving behind the salt dissolved in it. But the sea doesn't get any saltier: the salt content of seawater has maintained a fairly precise consistency of 35 parts per thousand for 250 million years. Where does all the salt go? Partly it is absorbed into the bones and shells of sea organisms; some of it is buried with the sediment on the seabed; some is washed onto the shore by wind and waves, and forms a crust; and some is left behind on the land when sea levels change – such as when the seabed rises, or the amount of water decreases.

Why is the sea saltier at the Equator and fresher at the poles?

The sun is fiercer at the Equator, so the seawater evaporates faster, raising the salt content significantly. In higher latitudes, there is a great deal of precipitation, which dilutes the salty water.

Why is there life in seawater ice and not in freshwater ice?

In salt water only the water molecules freeze; the salt remains in a fluid solution and forms little channels in the ice, from 100 microns to one centimetre in diameter, in which diatoms (single-cell algae) are able to thrive. These channels do not form in fresh water ice, so there is no room for life to subsist.

Why is the sky blue?

'If you look through the smoke of a fire made of dry wood towards a dark background, the smoke appears blue. That is why the sky looks blue, because there is darkness behind it,' wrote Leonardo da

Vinci in his notebook. He was the first to tackle the question of how the sky acquires its colour. He concluded that the more distant the object, and the more atmosphere the observer's gaze had to penetrate, the more the object lost its natural colour and appeared to be blue. That is why distant mountains are also blue.

In the next few centuries the best scientists racked their brains over this question: de Saussure, Newton, Bouguer, Tyndall, Arago, Lorenz. But then John William Strutt, later Lord Rayleigh, established that light is scattered as it passes through the atmosphere. Part of the light is 'thrown off course' by dust particles in the air. Blue light is much more subject to scattering than red light, which means that for the most part it is the blue component of the scattered light that reaches the Earth. If we look at the sky, then, we only see blue. The molecules in the air act as a kind of 'pigment' to render the earth's atmosphere visible.

Why is the sea blue?

The blue of the sea is caused by the light that falls upon it. Clear water absorbs every colour except blue. The blue colour is reflected.

Why is it green for starboard?

1834 – Ships of the City of Dublin Steamship Company were equipped with white masthead and starboard lights, and red navigation lights on the port side.

1836 – The P & O Company of Southampton had a different arrangement: green for port, green and red for starboard.

1847 – The British Admiralty ordained that starboard was to be green and port red.

1853 – The Prussian Ministry of Trade prescribed the British lighting rules for all steamers.

1858 – France, Austria-Hungary, and the North German seaboard
  countries also signed up.
1889 – 27 other seafaring nations followed suit, and adopted international maritime regulations.
The colour code applies to lighthouses, too – those on the right of har-
bours (when approaching from the sea) and on river or canal banks
are always green-striped.

What can fish hear?

Only 75 species of fish out of 25,000 can hear. Here are five exam-
ples, with their frequency ranges:

American shad	20–180,000Hz
Goldfish	30–3000Hz
Atlantic salmon	20–500Hz
Tuna	100–500Hz
Atlantic cod	100–300Hz

How far is the horizon?

Lying on the beach, eye height 20cm from the ground	1.75 km
A child (1 m) standing on the beach	3.92 km
An adult (1.70 m) standing on the beach	4.66 km
Standing on a sand dune (5 m)	7.98 km
From an upstairs window (10 m)	11.29 km
From the upper deck of a ship (30 m)	19.55 km
From the top of Salisbury Cathedral (123 m)	43.51 km
From a height of 500m	79.83 km
From a height of 1000m	112.89 km
From the top of Ben Nevis (1344 m)	143.84 km
From Montblanc (4807 m)	247.52 km
From the summit of Mount Everest (8848 m)	335.74 km
From a flying altitude of 11,000 m	374.43 km

Republic of Minerva The multimillionaire Michael Oliver built an artificial island on a shoal in the South Pacific that was covered by a metre of water even at low tide. He founded a republic and promised all its citizens a truly paradisiacal way of life: everything was permitted as long as it did not harm others; there would be no taxes or government. His neighbour, the King of Tonga, obviously feared the contagion of anarchy, and in 1972 he moved in with his troops to put an end to this state of affairs. But Michael Oliver did not abandon his dream of a free island life, and by the following year he was trying to persuade the population of Abaco in the Bahamas of its virtues. When that too failed, he started a new independence movement on the island of Santo, which belonged to Vanuatu; it foundered in a confusion of bizarre election campaigns, attempted coups and general chaos. But Michael has not given up; it is reported that he is currently looking for converts on the Isle of Man, and on the islands of Santo and Aurora in the Azores.

Seborga This territory in Liguria, Italy, is a land-locked island of self-declared independence. It is situated close to the French border and measures fourteen square kilometres, and was somehow left off the register when the Prince of Sardinia made a land purchase in 1729. It was not listed during the unification of Italy, either. Seborga declared its independence in 1996 – though no one seemed very concerned about it.

Republic of Morat-Songhrati The Spratly Islands in the South China Sea consist of 500 to 600 coral reefs, atolls and needle-like rocky outcrops. James George Meads discovered them in 1877, and declared himself King James I of his 'Kingdom of Humanity'. As a sailor-king, he envisaged a sanctuary for the downtrodden and persecuted of the world. Until the outbreak of the Second World War, he and his descendants ruled peacefully and successfully over a population of about 2000 subjects. But then the French and the Japanese annexed the quirky kingdom. A neighbouring sultan intervened, a struggle for power broke out among the population, and a certain Christopher Schneider declared a republic. Together with Morton F. Meads, a descendant of the sailor king, he bought the islands back from the sultan, and tried to have it recognized as an independent nation under his flag, bearing the yin-yang symbol. But Schneider and his entire cabinet were drowned in a storm. Vietnam, Taiwan, China, the Philippines and Malaysia occupied various islands, and Meads left to set up his government-in-exile in Australia, from where he is still fighting to regain his realm.

Kalayaan After the Second World War, Japan renounced the Spratly Islands, and the Philippines laid claim to them, though without taking any further steps. So Thomas Cloma, a Filipino lawyer and owner of a small fishing fleet, seized the opportunity and in 1956 founded his 'Freedom Land' on 53 rocks rising precipitously from the sea. Cloma enjoyed good relations with the Philippine state, which left him alone. When Taiwan later claimed some of the reefs and occupied them, the rump of Kalaayan formally joined the Philippines.

Number of boats registered in Florida

1976	447,000
1980	500,000
1984	640,000
1988	700,000
1992	700,000
1996	750,000
1999	830,000

Number of manatees killed in the mangroves by Floridian boats

~

```
1976 . . . . . . . . . . . . . . . . . . . . . . . . . 10
1980 . . . . . . . . . . . . . . . . . . . . . . . . . 15
1984 . . . . . . . . . . . . . . . . . . . . . . . . . 32
1988 . . . . . . . . . . . . . . . . . . . . . . . . . 40
1992 . . . . . . . . . . . . . . . . . . . . . . . . . 33
1996 . . . . . . . . . . . . . . . . . . . . . . . . . 51
1999 . . . . . . . . . . . . . . . . . . . . . . . . . 82
```

Ghosts on the *Queen Mary*

~

≈ *Door 13* in the engine room: trying
to escape from a fire, a seventeen-year-old
man was crushed there, and can still be
heard knocking.
≈ In an empty dressing room by the
swimming pool: steps have been
heard, and footprints seen.
≈ In the *entrance hall*: a
woman in white can be
seen wandering about.
≈ *Strange smells* were
reported during
the Second
World
War.

Plimsoll line

In 1870 Samuel Plimsoll MP (1824–98) began his campaign in the British Parliament against the overloading of ships and wrote a book called *Our Seamen*. Overloaded and badly maintained 'coffin' ships caused countless losses before load lines became compulsory; in 1873–4, for example, around the coast of the UK, 411 ships sank with the loss of 506 lives. Some of these losses were caused by insurance frauds, and attempts were made in Britain to enforce loading standards including the ship insurer Lloyd's Register in 1835.

In 1872 a Royal Commission on Unseaworthy Ships was set up and eventually the Plimsoll mark was made compulsory in Britain by the Merchant Shipping Act of 1876. In 1906, foreign ships were also required to carry a load line if they visited British ports.

In 1876, Fred Albert wrote the popular song 'A Cheer for Plimsoll':

> So a cheer for Samuel Plimsoll and let your voices blend
> In praise of one who surely has proved the sailors' friend
> Our tars upon the ocean he struggles to defend
> Success to Samuel Plimsoll, for he's the sailors' friend.
>
> There was a time when greed and crime did cruelly prevail
> And rotten ships were sent on trips to founder in the gale
> When worthless cargoes well-insured would to the bottom go
> And sailors' lives were sacrificed that men might wealthy grow.
>
> For many a boat that scarce could float was sent to dare the wave
> 'til Plimsoll wrote his book of notes our seamen's lives to save;

His enemies then tried to prove that pictures false he drew
But with English pluck to his task he stuck, a task he deemed
    so true.

Anson's scurvy-blighted voyage

Commodore George Anson's round-the-world voyage of 1740–4
was blighted by scurvy and starvation which killed almost 1400 of
the 1900 sailors who had set out from Spithead in six men-of-war.

The symptoms of scurvy are almost too ghastly to repeat, but
Anson lists them as follows: large discoloured spots over the whole
body, swollen legs, putrid gums, an extraordinary lassitude, strange
dejection of spirits, dreadful terrors, putrid fevers, pleurisies, jaun-
dice, violent rheumatic pains, ulcers, healed wounds re-opening
and necrotized flesh.

Dr James Lind of Haslar Hospital hit on the solution in the 1740s
(though too late for Anson's crew): more fresh food, preferably raw,
and lemon juice. The Admiralty, almost half a century later, issued
lime juice to sailors, but this was nowhere near as effective as lemon
juice. The error apparently arose because the West Indians called
a lemon a lime.

Incredibly, in June 1743, with only the flagship *Centurion* left, Anson
intercepted and captured the fabulous Spanish treasure ship
Covadonga laden with 1,313,843 pieces of eight and 35,682 oz of
silver and plate. This was one of the most valuable treasures ever
seized by an English ship. It required 32 wagons to transport it to
the Tower of London on Anson's return in July 1744.

MSC *Napoli* runs aground

In January 2007 the 62,000-tonne MSC *Napoli* container ship grounded off Sidmouth, east Devon. The ship was carrying 3,500 tonnes of fuel oil and 2,318 containers, 103 of which went overboard. The beaching of about 50 containers at Branscombe sparked a two-day looting frenzy.

Among the beach booty were: sunglasses, Timberland boots, teddy bears, perfume, car batteries, bibles, BMW gearboxes and other spare parts, 50 new BMW K1200 GT motorbikes, Nike trainers, tennis balls, nappies, a 4x4 vehicle, hundreds of barrels of wine, dog food, cosmetics, oil paintings by the Russian artist Aidir Chusainov and a wooden crate carrying £130,000 worth of family heirlooms. Thousands of people picked over the wreckage on the beach amidst extraordinary scenes reminiscent of the 1949 Ealing comedy *Whisky Galore*. Branscombe and all roads leading to the beach had to be closed by the police to all visitor traffic.

Officially, anyone who finds unclaimed wreckage in territorial waters must inform the Receiver of Wreck or risk a fine of up to £2,500.

10 Best sea films

The Bedford Incident (Columbia, 1965) Director: James B. Harris. Cast: Richard Widmark, Sidney Poitier, Eric Portman. Based on novel by Mark Rascovich.
Tense cold-war destroyer versus U-boat drama with hints of Herman Melville's Moby Dick. *At all costs avoid discovering the ending before you watch it.*

The Cruel Sea (Ealing, 1953) Director: Charles Frend. Cast: Jack Hawkins, Donald Sinden, Stanley Baker, Denholm Elliott. Based on novel by Nicholas Monsarrat.

Realistic, gritty re-creation of World War II Atlantic convoys with no holds barred.

Jaws (Universal, 1975) Director: Steven Spielberg. Cast: Roy Scheider, Richard Dreyfuss, Robert Shaw. Based on novel by Peter Benchley.
Shocking great white shark terrorizes coastal resort town, and most audiences.

Dead Calm (Warner, 1989) Director: Phillip Noyce. Cast: Sam Neill, Nichole Kidman, Billy Zane. Based on novel by Charles Williams.
Chilling, nightmare yacht voyage in the Pacific. This could put deep-sea yacht crews off for life.

The Last Voyage (MGM, 1960) Director: Andrew Stone. Cast: Robert Stack, Dorothy Malone, Woody Strode, George Sanders, Edmond O'Brien. Screenplay: Andrew Stone.
Luxury liner SS Ile de France, *soon to be scrapped, becomes the SS* Claridon *and is used in this gripping sinking ship drama. Will trapped Dorothy Malone survive?*

Master and Commander: The Far Side of the World (Fox, 2003) Director: Peter Weir. Cast: Russell Crowe, Paul Bettany, James D'Arcy. Based on novels by Patrick O'Brian.
Captain Jack Aubrey and his gallant English sailors hunt the wily French frigate Acheron.

Das Boot (Columbia, 1981) Director: Wolfgang Petersen. Cast: Jurgen Prochnow, Herbert Gronemeyer. Based on novel by Lothar-Günther Buchheim.
The anti-war story of a German U-boat on a convoy-hunting mission in the North Atlantic.

Titanic (Paramount/Fox, 1997) Director: James Cameron. Cast: Kate Winslet, Leonardo di Caprio, Billy Zane. Screenplay: James Cameron.
Spellbinding, romantic disaster film with spectacular special effects.

Lifeboat (Fox, 1944) Director: Alfred Hitchcock. Cast: Tallulah Bankhead, William Bendix, Mary Anderson, Henry Hull. Based on novel by John Steinbeck.
World War II film about American and British survivors of a U-boat attack stuck in a lifeboat with an enigmatic German sailor.

San Demetrio London (Ealing, 1943) Director: Charles Frend. Cast: Walter Fitzgerald, Mervyn Johns, Robert Beatty, Ralph Michael, Gordon Jackson. Screenplay: Charles Frend & Robert Hamer.
Heroic incident from convoy HX-84 re-created in a tribute to Merchant Navy crews.

Viva Riva

Older models from the Italian Riva boatyard are reckoned to be the most exclusive motorboats in the world. Up to 1996, 4292 mahogany boats had been manufactured. Such is their rarity that the prices given are only estimates; prospective buyers should reckon on price variations of up to +/- 30%.

1950–7	137 *Scioattolo*		(Outboard)
1952–7	119 *Sebino*		
1946–55	40 *Corsaro*		
1952–64	426 *Florida*	5.59 m,	40,000 euros
1953–65	711 *Florida Super*	6.27 m,	40,000 euros
1950–72	804 *Ariston*	6.50 m,	90,000 euros
1956–8	19 *Ariston Cadillac*		
1960–74	181 *Ariston Super*		
1950–66	181 *Tritone*	8.02 m,	160,000 euros

1956–60 .... 10 *Tritone Cadillac*...........................

1960–3 ..... 21 *Tritone Super*...........................

1962–72..... 228 *Aquarama*.......... 8.14 m, 200,000 euros

1963–71..... 203 *Aquarama Super*...... 8.75 m, 250,000 euros

1972–96 .... 277 *Aquarama Special*..... 8.75 m, 350,000 euros

1966–72 .... 626 *Junior*................. 5.70 m, 30,000 euros

1969–79 .... 264 *Olympic*.............. 6.55 m, 50,000 euros

The shipping forecast

~

The forecast is read out on BBC Radio 4 LW on 1515 m (198kHz) – (some transmissions on VHF) at 00.48 hrs, 05.20 hrs, 12.00 hrs and 17.55 hrs. Gale warnings are given as necessary during the day. The UK inshore (up to 12 miles offshore) forecast is read on BBC Radio 4 at 00.48 hrs and on BBC Radio 3 at 05.35 hrs.

The music played before the shipping forecast is 'Sailing By', composed by Ronald Binge (1910–79), one of the most respected and successful English composers of his generation. Some years ago the BBC planned to ditch 'Sailing By' in favour of something more 'modern' but an outcry from devotees forced a change of heart.

Those terms explained:
Gale warnings

Gale	Force 8 winds (34–40 knots) or gusts reaching 43–51 knots
Severe Gale	Force 9 winds (41–47 knots) or gusts reaching 52–60 knots
Storm	Force 10 winds (48–55 knots) or gusts reaching 61–68 knots
Violent Storm	Force 11 winds (56–63 knots) or gusts of 69 knots or more
Hurricane Force	Force 12 winds (64 knots or more)

Imminent	Expected within 6 hours of time of issue
Soon	Expected within 6 to 12 hours of time of issue
Later	Expected more than 12 hours from time of issue

Visibility

Fog	Visibility less than 1000 metres
Poor	Visibility between 1000 metres & 2 nautical miles
Moderate	Visibility between 2 & 5 nautical miles
Good	Visibility more than 5 nautical miles

Movement of pressure systems

Slowly	Moving at less than 15 knots
Steadily	Moving at 15 to 25 knots
Rather quickly	Moving at 25 to 35 knots
Rapidly	Moving at 35 to 45 knots
Very rapidly	Moving at more than 45 knots

The hunt for the USS *Scorpion* and the sandheap theorem

On 21 May 1967 the USS *Scorpion*, one of the US Navy's six atomic-powered submarines, was 50 sea miles from the Azores when it sent out a radio message. It was the last sign of life ever received from the ship. Six days later the boat was officially declared lost. But where had it gone down? At its normal rate of progress, in six days the *Scorpion* could have reached any point in the North Atlantic. The mathematician, John Cramer, was called in to help with the search. He applied a new technique, Bayesian probability theory, which had already proved successful in locating a missing hydrogen bomb. He asked experts in various fields – the military, accident specialists, ocean-current researchers – to give him their best estimate of where the submarine might be located, offering a bottle of Chivas Regal for the ones who turned out closest to the mark. Then Cramer reduced the widely varying responses to a single calculation, balancing out the various uncertainty factors; and lo and behold – the mean value of the positions suggested by the experts led the Americans to their submarine. It lies 3000 metres down, 700 sea miles (740 kilometres) south-west of the Azores. This so-called sandheap theorem has proved very useful since. What is important is to weight individual hypotheses: expert against experts, and layman against layman.

A sea of books

Bible *The Flood*
Homer *Odyssey*
Virgil *Aeneid*
1001 Nights *Sindbad the Sailor*
Luis de Camões *Lusiads*
Daniel Defoe *Robinson Crusoe*
James Fenimore Cooper *The Red Rover*
Edgar Allan Poe *The Narrative of Arthur Gordon Pym*
Charles Darwin *The Voyage of the Beagle*
Herman Melville *Moby Dick*
Victor Hugo *Les travailleurs de la mer*
Jules Verne *20,000 Leagues Under the Sea*
Robert Louis Stevenson *Treasure Island*
Pierre Loti *Pêcheur d'Islande*

Erskine Childers *The Riddle of the Sands*
Joseph Conrad *Nostromo*
Jack London *The Sea Wolf*
James Joyce *Ulysses*
B. Traven *The Death Ship*
E. C. Segar *Popeye*
Richard Hughes *A High Wind in Jamaica*
Ernest Hemingway *The Old Man and the Sea*
Ignacio Aldecoa *Gran Sol*
Stanisław Lem *Solaris*
Sten Nadolny *The Discovery of Slowness*

Arthur Ransome's best seafaring book for children of all ages – *We Didn't Mean to Go to Sea*, first published in 1937 – was inspired by the yacht *Nancy Blackett*, which he bought in 1934 for £525.

The boat, which had been built in 1931, was mostly sailed around the east coast of England, but made one trip to Holland to provide background information for the book. In 1938 Ransome sold *Nancy Blackett* under some pressure from his wife, who wanted a boat with a larger galley.

In *We Didn't Mean to Go to Sea* most of the action takes place on board *Goblin* (a thinly disguised *Nancy Blackett*), which plays the leading role in this dramatic story. After the Ransomes parted with *Nancy Blackett* she had five other owners until, in 1988, she was discovered in Scarborough Harbour in very poor condition. Fortunately, she was taken on as a restoration project and in 1996 ownership passed to the Nancy Blackett Trust, dedicated to the preservation of this classic yacht in full sailing order.

Acronyms

LAMBADA (Large Scale Atmospheric Moisture Balance of Amazonia): *Data-gathering project studying energy, heat and moisture in the Amazon Basin*

MAMBO (Mediterranean Association of Marine Biological Oceanography)

SAMBA Sub-Antarctic Motions in the Brazil BAsin: *How water travels from the Antarctic to the equator*

ELOISE (European Land Ocean Interaction Studies): *the EU's project on coastal regions*

BOAT (Bulletin Océan Atlantique Tropical): *a French scientific publication*

FROST (First Observing Study of the Toposphere): *weather observation in the Antarctic*

OCTOPUS (Ocean Colour Techniques for Observation, Processing and Utilization Systems): *identifying environmental change by means of observing changes in the colour of the sea*

PROTEUS (PROfile TElemetry of Upper ocean currentS): *deep-sea data gathered by satellite*

REMUS : REchnergestütztes Maritimes Unfallmanagement System. *Computer-supported maritime accident management system*

ALEXIS (Array of Low Energy X-Ray Imaging Sensors): *Extreme ultraviolet and low-energy X-ray space-imaging telescope*

CAESAR (Coordinating and Educating Search and Rescue): *sea-rescue training*

DORIS (Doppler Orbitography and Radiopositioning Integrated by Satellite): *Doppler satellite tracking system developed for precise orbit determination and precise ground location*

TED (Turtle Excluder Device): *device to help sea turtles escape fishing nets.*

GIN: Greenland/Iceland/Norway.

SCUBA : Self-Contained Underwater Breathing Apparatus.

The Chinese mitten crab

A member of the *grapsidae* family; the claws of the male are covered in patches of hair. This short-tailed or edible crab was introduced in ballast water on Chinese merchant ships around 1910, and has since spread via the Rhine and the Elbe as far as Basel and Prague, as well as establishing itself elsewhere in western Europe, including the UK. This busy predator migrates upriver over a number of years, surmounting apparently insuperable obstacles, and making long marches (sideways) across land, sometimes covering an astonishing 12 kilometres in a day.

In late summer the Chinese mitten crab returns to the sea to mate. It lives in fresh or brackish water, burrows into muddy river banks, and steals from fishing creels. For this reason it is not very popular with fish farmers. However, over the last hundred years the native fauna have come to terms with these immigrants. Although it tastes good (boil for 15 minutes), the crustacean, about 30 centimetres long, looks unattractive with its olive-green flesh, and is eaten only in Chinese restaurants – by the staff.

Great Barrier Reef faces bleak future

Australia's Great Barrier Reef could become a bleached coral wasteland within 25 years if global warming continues to create warmer and more acidic seas. The 1900 km-long reef off the Queensland coast covers over 347,000 sq km and is home to 1500 species of fish, 4000 types of mollusc and 400 different types of coral. Bleaching occurs when the water reaches a critical temperature and the tiny living polyps that make up the coral are killed, leaving behind the white limestone skeleton of the reef. The reef, the world's largest living organism, is a major attraction in Australia's £1.8 billion-a-year tourist industry.

Cold-water coral reefs

Sula-Reef, Røst, Tissler Riff (Norway)
Darwin Mounds (Scotland)

North Sea (and North East Atlantic) Coral

Devonshire cup coral (*Caryophyllia smithii*)
Tuft coral (*Lophelia pertusa*)
Kreiselkoralle (*Stephanotrochus moseleyanus*)
Scarlet and gold star coral (*Balanophyllia regia*)
Dead man's fingers (*Alcyonium digitatum*)

Dead man's fingers

~

The soft or eight-branched coral (Alcyonaria) occurs on the Atlantic coast from Ireland to Portugal, as well as in the North Sea and occasionally the Baltic. As it does not form a skeleton, it does not build up coral reefs. Soft coral likes dark, cool, flowing water down to 50 metres; it clings to rocks, and even to larger species of crab, and feeds on drifting plankton. It builds up male or female, and sometimes hermaphrodite, colonies, releasing sperm or eggs in December and January. These reddish, white or brownish colonies survive for up to 20 years and grow to 30 centimetres. If you fish them up out of the water their sausage-shaped excrescences are like grasping a cold, dead hand.

The great auk

~

The great auk was the only flightless bird native to the northern hemisphere. It looked like a penguin – and was in fact the original penguin, for when the Spaniards discovered similar birds in the Pacific, they named them after the bird they knew in the far north. Its original habitat was the coast on both sides of the Atlantic, and it survived on inaccessible islands in the North Atlantic, until the cod fishermen wiped it out there too. There was no need to waste valuable ammunition hunting it, a fatal attribute it shared with the unfortunate dodo of Mauritius. Its feathers provided down, while its fat supplied fuel, just as penguins in the Antarctic served as substitutes for firewood. By the time nineteenth-century naturalists became keen to encounter the fabulous bird, it was already almost extinct. The last colony on Iceland was purloined by locals and sold to naturalists. In May 1834 the last of its kind was captured, half-starved, and kept as a domestic pet. The big, lonely bird was spotted a few times in the following years on mist-shrouded islands near the Arctic Circle, a mirage from the past. The great auk is no more.

The wingless flies of the Kerguelen Islands

The Kerguelens are a very windy archipelago in the Southern Ocean, quite close to Antarctica. Their only interesting feature being a fly without wings. While such insect mutations normally die out, on the Kerguelens only the *Calycopteryx moseley* can survive; all the rest get blown away.

The albatross

Thanks to its enormous wingspan, when gliding the albatross loses only one metre in altitude for 22 metres of flight. In the right weather conditions, this is what enables it to cover 1,000 kilometres a day. It is also a good diver, and can live to the age of about fifty. Fishermen used to catch it out of the air with rods, but nowadays 100,000 a year perish as by-catch of the big trawlers.

What type of albatross did the Ancient Mariner shoot?

In 'The Rime of the Ancient Mariner' by Samuel Taylor Coleridge (1772-1834), first published in *Lyrical Ballads* in 1798, the killing of an albatross leads to grotesque and protracted horror for all on board a doomed ship. The idea was probably inspired by Captain

George Shelvocke's *A Voyage round the world* published in 1726, which had been drawn to Coleridge's attention by William Wordsworth. Shelvocke's book describes a 'disconsolate black albatross' which followed his ship during constant foul weather. This bird was considered to be an evil omen and it was shot in the hope of encouraging fair winds.

In Coleridge's poem the ancient mariner kills an albatross with his crossbow in a moment of madness and the ship and its crew are destined for a sticky end. The mariner is blamed and the albatross is hung about his neck.

But what type of albatross was it? The most familiar type is the wandering albatross (*Diomedea exulans*), a huge white bird with jet-black wingtips, weighing 7–11 kilos with a wingspan of 3–4 metres. This bird was too large and heavy to have been hung around anyone's neck. However, the dark-mantled Sooty Albatross (*Phoebetria fusca*) with distinctive black plumage is much smaller – about the size of a goose – with a wingspan of about 2 metres and the most likely victim of the mariner. Coleridge does not specify the colour of the bird but it is very likely that he had in mind the 'black albatross' mentioned in Shelvocke's book.

Phoebetria fusca nests on islands in the South Atlantic (Tristan da Cunha & Gough Island) and islands in the South Indian Ocean (the Crozet Islands to the Kerguelen Islands). At sea they roam from South America to Australia.

The frigatebird

~

The fastest sea bird in the world (up to 400 km/h in a dive) has a wingspan of 2.3 metres, and weighs 1.2 kilos. In the South Seas it is used for fishing and as a kind of 'carrier pigeon'.

Zino's petrel

~

The rarest seabird in the world was only identified in 1903, and in the 1950s it was thought to have died out. In the 80s, shell fragments from the eggs (eaten by rats) of this ground-breeding bird were found. A rescue programme was set up, and today there are about 70 breeding pairs of Zino's petrel on Madeira.

Pigeons save RAF crews

~

Four pigeons of the National Pigeon Service saved aircrew from the sea during World War II:

'Winkie'	February 1942
'Dutch Coast'	April 1942
'Tyke/George'	June 1943
'White Vision'	October 1943

These heroic pigeons received their Dickin medals on 2 December 1943. The citation for 'Winkie' reads:
'For delivering a message under exceptionally difficult circumstances and so contributing to the rescue of an Air Crew while serving with the RAF in February 1942.'

The 42 Squadron Beaufort Torpedo Bomber crew that 'Winkie' saved were pilot Sqn Ldr W. H. Cliff, navigator F/O McDonald, Wop/AGs P/O Tessier and Sgt Venn. They were returning from an operational hunt for German shipping off the Norwegian coast when an engine fire caused the plane to crash into the ice-cold North Sea about 150 miles from their base at Leuchars, Scotland. The crew boarded the rubber dinghy together with one pigeon basket containing 'Stinkie'. 'Winkie' was believed to have drowned. A short message with their estimated position was attached to

'Stinkie's' leg and the bird was launched in the general direction of its loft at Broughty Ferry near Dundee. A brief radio SOS and call sign had been sent but the pigeon was their best hope.

Air searches over the North Sea for the missing crew drew a blank and hope was fading. Meanwhile, a bedraggled bird with its feathers clogged with oil appeared in the pigeon loft of James Ross but there was no message. After much inspired guesswork it was decided that the pigeon must have hitched a ride on a passing oil tanker and spent the hours of darkness on board before struggling home in the early hours. The ship was identified and its track added to the calculations.

The search area was now moved south and a RNAF Hudson successfully located the dinghy, and the crew were picked up by high-speed launch. They were all suffering from exposure and frostbite and would probably not have survived another night at sea.

The crew returned to base and met James Ross who told them that it was 'Winkie' who saved them. 'Stinkie' had sadly not survived. There is a heart-warming photo of the four crew members with their feathered saviour, who lived until 1953.

The Dickin medal winner was stuffed and can now be seen at the Imperial War Museum.

The Dickin medal

This award came into being in 1943 at the suggestion of Maria Dickin, the founder of the People's Dispensary for Sick Animals (PDSA). During World War II she was inspired by the displays of bravery shown by animals and birds used on active service to introduce a medal to recognize their courageous efforts. Of the 53 Dickin medals awarded during the war, 32 were won by pigeons.

UK marine life

over 44,000 species, from plankton to whales
over 330 species of fish
25 breeding seabird species
8 million coastal birds
90% of the global population of Manx shearwaters
95% of the EU's grey seal population

UK fisheries minister Ben Bradshaw said that the fight to preserve marine life was 'the next biggest environmental challenge the world faces after climate change'.

Know your crustaceans

≈ shrimps and prawns are Natantia (swimmers).
≈langoustines (aka Dublin Bay prawns, Norway lobster) and lobsters are Reptantia (crawlers).
≈ all of the above belong to the *Decapod* (ten-legged) order of crustaceans.

Swallow, don't chew

Source: I.F.O.C.E.
(International Federation of Competitive Eating)
≈ *4.4 kg of lobster* – polished off by the redoutable Sonya Thomas in 12 mins (2005)
≈ *2.055 kg prawns* – consumed by Charles Hardy in 12 mins
≈ *331 crayfish* – devoured by Chris Hendrix in 12 mins
≈ *552 oysters* swallowed by Sonya Thomas (again) in 2005 in 10 mins

Recipe for cream of sea urchin

Ingredients:

4 sea urchins	white wine	one shallot
butter	two egg yolks	crème fraîche
		fish stock

Preparation:
Hold the sea urchin with a cloth in one hand, flat side upwards. Make a circular hole with scissors and remove the segment of shell and the guts. Strain the juice and catch it in a bowl. Spoon out the ovaries or gonads and tongues and put them aside. Warm up the butter, fry the chopped shallot until transparent, and then add wine. Reduce the sauce by half, add the fish stock and sea urchin juice, then continue to boil it down. Mix in the egg yolks and the crème fraîche, and add to the warm sauce, stirring continuously. Strain the thickened sauce and then beat it with a whisk. To prevent the egg yolk from solidifying, be very careful not to let the sauce boil. Heat carefully, continuing to beat. Season to taste and add sea urchin tongues just before serving.

Stuffed squid with tomato sauce

To prepare the sauce:
Sweat some shallots in olive oil, add a tin of peeled tomatoes, and simmer gently. Pour in vegetable stock or broth. Many chefs add a dash of wine or a whole clove of garlic. A pinch of sugar is also recommended.

To stuff the squid:
Wash and clean out squid tubes (remove the head and attached tentacles, pull out the pen – the quill-shaped bone – from the mantle cavity). There are several options for the stuffing: you can

chop the tentacles up finely and knead together with white bread and coriander; or make a purée of chestnuts, raisins and Parma ham; or a paste of pine kernels, anchovies and hard-boiled eggs; or make little balls out of crabmeat, cooked rice and parsley. Whatever you choose, push the stuffing into the squid tubes. You can close them off with a toothpick, but that is not usually necessary. Place them in the simmering sugo for 20 minutes. Turn over once. Serve with ciabatta, salad and white wine.

Recipe for sea cucumber

Wash the sea cucumber thoroughly, removing all the dirt from between the spines. Soften and clean the sea cucumber by leaving it in cold water for three days (change the water frequently).

Then cut the cucumber open lengthwise. Take out the gut, rinse out thoroughly, and cut into thin strips, four centimetres long. Blanch for about three minutes in boiling water.

Cut up about 100 grams each of cucumbers, carrots and leeks into sticks. Pour peanut oil into a hot wok or pan, and briefly stir-fry 40 grams of chilli bean paste and 20 grams of ginger cut into sticks. Pour on a dash of rice wine (or sherry) and 400 millilitres of chicken broth, and bring to the boil. Skim the broth.

Add the sea cucumbers, cover the pan with a lid, and cook on a low heat for about ten minutes. Meanwhile, bring some water to the boil in a second pan, put in the carrots, cucumbers and leeks, and blanch for about a minute. Serve together.

Don't mess with a sea cucumber

A sea cucumber is not a vegetable, but a coelenterate (hollow-bodied animal). It breathes through its anus. Small fish enter through its body opening to graze on the gut. If a holothurian (sea cucumber) feels threatened, it shoots out parts of its intestines through its anus at the attacker; they grow back afterwards. The animals have strange names, too: pineapple sea cucumber, apple sea cucumber and donkey dung sea cucumber.

'The boy stood on the burning deck'

On 1 August 1798 a British naval squadron under the command of Rear Admiral Horatio Nelson caught the French fleet at anchor in Aboukir Bay near Alexandria. The ensuing engagement became known as the Battle of the Nile and Nelson's crushing victory meant that Napoleon's army was cut off in Egypt and was forced to surrender by a British force in 1801.

During the battle *L'Orient*, the French flagship of Admiral Paul de Bruey, under the command of Captain Luc-Julien-Joseph Casabianca, was set ablaze by English broadsides. The French guns were abandoned and it was clear that the ship was doomed. The English sailors were amazed to see a boy standing alone at his post surrounded by flames on the burning deck. Soon afterwards the fire reached *L'Orient*'s powder magazine and an enormous explosion destroyed the ship and killed twelve-year-old Giocante Casabianca, his father and most of the crew. Every ship in the vicinity was showered with debris which included a large piece of *L'Orient*'s mainmast which was presented to Nelson. After his death at Trafalgar in 1805 wood from this trophy was used to construct Nelson's coffin.

The poem 'Casabianca' by Felicia Hemans (1793–1835) begins with the immortal words: 'The boy stood on the burning deck' and recounts how the 'creature of heroic blood' waited in vain at his post for his father's order to release him from his duty. The poem was first published in the *Monthly Magazine* for August 1826 and became a firm favourite for many years, although in our more cynical age, it is more likely to be the subject of parody than admiration.

The world's biggest container ship

Emma Maersk was built at the Odense Steel Shipyard in Denmark and is powered by a Wartsila-Sulzer RTA 96-C engine, currently the world's largest single diesel unit, weighing 2300 tons.

Capacity	11,000 TEU (Twenty-foot equivalent units) containers are either 20 or 40 feet long.
Length	397 metres
Beam	57 metres
Gross tonnage	156,907
Maximum draft	16.5 metres
Engines	110,000 h.p.
Crew	13
Annual mileage	170,000 nautical miles

If all the containers on board were lined up end to end, they would stretch 67.6 km.

Freighters lose about 10,000 containers every year. Many split open when they hit the water and release their contents; this is often a disaster for the environment – but also a unique opportunity for research into ocean currents.

In 1990, 80,000 Nike trainers were lost overboard from the *Hansa Carrier*, about 700 kilometres south of the Aleutian Islands. Over 2,000 single shoes were picked up on the beaches of British Columbia, Oregon and California. Anyone lucky enough to find a matching pair will have discovered that they were still wearable. They had not suffered from their long soaking.

On 12.12.2002, a container with 33,000 basketball shoes (Nike again) slipped into the sea during a storm off Cape Mendocino in California. The fleet of sports shoes is making its way north at a rate of 360 miles per month.

In the same storm 17,000 tins of noodles (Chow Mein) went overboard. They, too, have been collected from beaches, and it seems that some are still quite edible.

Since 10.1.1992, 28,000 plastic ducks, frogs, tortoises and beavers have been drifting around the oceans. After being shipwrecked in the Pacific, the colourful fleet sailed as far as Alaska, and from there they were swept with the ocean currents around the world. Many have been recovered in the meantime, but four of the plastic animals were sighted as late as 2005. The plastic ducks have by now gained global fame, and collectors are willing to pay three figure sums apiece for them.

On 13.2.1997 the *Tokio Express* lost three containers 32 kilometres from Land's End; precisely 4,756,940 Lego bricks poured into the sea. In 14 months they had drifted to Spain, then on to the Canaries, from where they were blown like Columbus by the trade winds in the direction of Florida. Oceanographers calculate that the plastic pieces will reach Alaska some time around 2012.

In 1994, the *Hyundai Seattle* lost 34,000 *ice-hockey gloves* overboard in the Pacific.

Eggs of the elephant bird, which became extinct three hundred years ago, were washed up in Australia – 10,000 miles from the coast of Madagascar, where the giant flightless bird had once laid them.

Messages in bottles

'Please help us,
President Bru,
or we are lost.
Richard Dresel.'
In 1939, Jewish
refugees from
Nazi Germany on
board the *St Louis*
were wandering the seas.
No country would accept them,
and eventually the ship was forced to return to Germany. A message in a bottle was thrown overboard for every one of the 847 passengers, containing an appeal to Cuban President Fredrico Lerdado Bru to grant them entry to his country. One of these appeals was found by John Moore at an auction in Bath in 2003, where he bought a copy of the book *Voyage of the Damned*, which tells the story of the odyssey of the *St Louis*. To this day Zilla Coorsh, Richard Dresel's, daughter, has no idea how this message from a bottle ended up in the book.

In 1979 the Chinese dissident Wei Jingshen was arrested. Eleven years later news of his arrest turned up in a bottle on a beach in Vancouver.

The Japanese student Satoshi Ohta threw 750 MIBs (messages in bottles) into the sea in 1984 and 1985, in order to study the Kuroshio Current. 49 were found in Alaska.

'Dear Steve, the weather is great,' read Ron Hulse in a coconut he picked up on 29.2.1996 in Ice Bay, Alaska.

A Japanese soldier was left stranded on an island during the Second World War. Facing death, he wrote his wife a farewell message in a coconut. His widow received it 35 years later.

'Whereas I now lie near starvation and perish, I, Morris A. Taylor of Overland, Missouri, USA, do establish this as my last will and testament. To my brother Roy I leave and bequeath Dollar 10,000. The entire remainder of my estate I leave to my beloved wife Karen Houseman Ta[y]lor.' These last words were scratched into a tree trunk in 1840 on one of the islands lying off southern Florida; the trunk was found drifting near Cocoa Beach in 1964. A hurricane had probably torn up the tree and hurled it into the sea.

The probability of a message in a bottle being found lies between one per cent and fifteen per cent, according to a researcher into messages in a bottle, Curtis Ebbesmeyer.

'Whisky Galore'

At 7.40 am on 5 February 1941 the SS *Politician*, en route from Liverpool to Kingston, Jamaica, and New Orleans, ran aground on sandbanks off the Isle of Eriskay, Outer Hebrides, and became a legend. On board in number 5 hold were £3 million of Jamaican banknotes and 22,000 cases containing 260,000 bottles of first-class whisky, tax unpaid, destined for the American market.

The ship flooded and the crew were taken off by the Barra lifeboat. Captain Worthington reported to the insurers that he thought the ship was salvageable.

In 1941 whisky was almost unobtainable and temptation got the better of the islanders who soon descended on the ship to liberate the cargo before it disappeared. The local customs officer, Charles McColl, and the police had their work cut out trying to track down the scavengers who resorted to all sorts of tricks to conceal their booty. A number of islanders were fined and some even jailed as McColl persevered with his crusade. Estimates of the islanders' illicit haul of amber nectar vary between 24,000 and 84,000 bottles but of course no one at the time was concerned with keeping an accurate record.

The eventual salvage attempts were largely unsuccessful and it was reluctantly decided to leave the ship where she was. McColl was not to be beaten and he was granted permission to blow up the wreck. 'Dynamiting whisky. You wouldn't think they'd be men in the world so crazy as that!' said one of the islanders, Angus Campbell.

Author Compton MacKenzie was at the time living on the Isle of Barra and was inspired to write the novel *Whisky Galore*. The 1949 Ealing film version, written by the author with Angus McPhail, was directed by Alexander Mackendrick and became a worldwide hit and well-loved classic comedy.

The Isle of Eriskay's other claim to fame is that Bonnie Prince Charlie landed in July 1745 from France to go on to gather together the clans for the Jacobite rebellion.

John T. Randall (1905–84) and Harry A. H. Boot (1917–83) of the Physics Department of the University of Birmingham made the first cavity magnetron work on 21 February 1940. This vital break-through made it possible to produce short-wavelength centimetric radars which were small and light enough to be fitted to anti-U-boat aircraft and ships which could spot small targets such as a submarine's conning tower at a distance. The first operational use of the new centimetric radar at sea was in the Bay of Biscay on 1 March 1943. The U-boat eventually became an endangered species in large measure as a result of this crucial invention.

The first radars had aerials 100 m high and the research team at Birmingham University were asked to aim for a 10-cm wavelength radar with a correspondingly smaller aerial. The GEC Wembley Research Labs developed Randall and Boot's laboratory device and produced a high-power, pulsed, sealed valve suitable for opera-tional use in June 1940. One of these magnetrons was flown to the USA by the Tizard mission and put into full-scale development and mass production. By autumn 1944 the Allies relied increasingly on the remarkably effective new 3-cm radar sets operated on ships and aircraft.

The cavity magnetron helped change the outcome of World War II. It also made possible the microwave ovens found in kitchens around the world.

'The Wreck of the Deutschland' and the Kentish Knock

'The Wreck of the Deutschland' by Gerard Manley Hopkins (1844–89) was inspired by a moving account of the tragedy off the Kent coast in *The Times*. The *Illustrated London News* also reported it in its edition of 18 December 1875. Hopkins dedicated the poem: 'To the happy memory of the five Franciscan nuns exiles of the Falk Laws drowned between midnight and morning of December 7th 1875.'

The North German Lloyd Company ship *Deutschland*, carrying 200 passengers and crew, was destined for America but was driven onto the Kentish Knock sandbanks by bad weather. The five nuns from Westphalia had been driven from Germany by the Falk Laws which aimed to restrict and regulate the clergy. Opposition to these laws eventually compelled the government to change its policy. The graves of the nuns are to be found in St Patrick's Roman Catholic Cemetery in Leytonstone, East London.

'The London Array' windfarm of 270 turbines is planned for the 45-square-nautical mile area around the Kentish Knock and the Kentish Deep and is to be constructed between 2007 and 2011.

Phantom islands

Endless days at sea, storms, fog, and then suddenly land in sight … That is the experience of many discoverers. However, many previously undiscovered coasts are never found again, although they are indicated on maps. Are these cases of delusion? Or mistaken identity? Or have the islands disappeared? Phantom islands wander in all directions on maps, becoming smaller and finally evaporating altogether.

'It's true, I have heard it myself – and not just from one person, but from many seafarers I have sailed with – that when they pass through these zones they hear a great, confused and indistinguishable din of human voices, coming out of the sky and the mast-tops. It was like the noise you hear from a crowd in a market place. When they heard this noise, they knew they were not far from the Isle of Demons.' (André Thevet, 1555)

On this island Jean-François de la Rocque, Sieur de Roberval, responsible for an expedition in 1542 to colonize Canada for France, marooned his niece Marguerite de la Roque, who had been conducting a love affair with a young officer. The officer managed to jump ship and join Marguerite and her elderly nurse, also left behind on the island because of her indulgent attitude to the lovers' conduct. Praying together, they did battle against the terrible demons and other 'monsters, so abysmally and inexpressibly horrible, the spawn of Hell, howling in their frustrated rage'.

Marguerite became pregnant, and the demons cursed and howled all the more. The father, the nurse and the baby all died. Marguerite shot a bear 'as white as eggs', and after two years and five months she was rescued by a fishing boat and returned to France.

The *Isle of Demons* is probably Fishot Island in St Lawrence Bay. The demons were brooding gannets and great auks.

Frisland

The Italian Sir Nicolò Zeno took his ship through the Strait of Gibraltar, and sailed for a certain number of days, maintaining a northerly course, with the intention of visiting England and Flanders. But on the high seas he ran into a terrible storm, which tossed the ship around so unpredictably that he no longer knew his position. Then he finally sighted land. Unable to withstand the power of the storm any longer, the boat was thrown upon the shores of the island of Frisland. The crew and most of the cargo were saved. This was in the year 1380.

Large numbers of the islanders arrived armed to attack Nicolò and his men. The latter were still suffering badly from the effects of their plight at sea and from exhaustion. They had no idea where in the world they had been cast ashore. But just as they were about to be taken prisoner, a Prince Zichmni came to their aid, asking in Latin where they hailed from. Nicolò and Zichmni became friends, and went off together on privateering raids. The Prince rewarded the Italian handsomely, and the latter lived on Frisland for 14 years. During an expedition against Estonia the friends found themselves in a bad storm, and took refuge on Grisland, a large, uninhabited island. From there they explored Iceland, where hot water shoots out of the ground and volcanoes spew out lava.

Frisland was probably the Faroes.

The Island of Buss

On Frobisher's third Arctic expedition, one of his ships, a large fishing vessel called a busse, became unfit to continue. On its journey back to England, when it was 'south-east of Freseland', it came across 'a great Ilande ... which was never yet found before, and sayled three dayes along the coast, the land seeming to be fruitfull, full of woods, and a champion countree'. That was the first appearance of the mysterious Buss Island, discovered by James Newton in September 1578.

In 1606 and in 1671, the 60-mile-long island, at latitude 58 and between longitudes 27 and 31, was sighted again. From 1594 until well into the 18th century there was a general belief in its existence, but as no sailor ever saw it again, it was believed to have sunk after a volcanic eruption.

Buss Island was probably either an iceberg, or the coast of Greenland in the mist, or it was a few rocks, some mirages and a lot of imagination.

Also known as: Tir fo-Thuin – the land beneath the waves; Magh Mell – the land of truth; Hy na-Beatha – island of life; Tir-na-m-Buadha – land of virtue; Tir Tairngiri – land of promise; Terra Repromissionis – the promised land of saints; Ysola Brazir, Bracir, Hy-Breasil, Hy-Brasil.

The really strange thing about this island is not so much the Celtic legends about it, or the reports of sailors about fairy folk, silent old princes and mysterious animals, but rather its precise location in the West of Ireland, near Galway Bay. At least half a dozen ships failed in their quest to find *Brasil*, but none the less belief in this imaginary island with its fabulous wealth, the home of the gods, has persisted from 520 until today, long after its cartographic death in 1865.

'On a clear evening with a beautiful golden sunset, exactly at the moment when the sun sank into the sea, a dark island appeared far out to sea, but not on the horizon. It had two hills, of which one was wooded. On a plain, towers rose, and columns of smoke curled upwards,' wrote the poetically gifted T. J. Westropp as late as 1872.

The island of *Brasil* really may have sunk; on its supposed site today, shallow-water mussels are found.

Gulf stream

~

The blockbuster disaster film *The Day after Tomorrow* (2004) is based on the idea that global warming could cause so much polar ice to melt that the salinity of the North Atlantic Drift is drastically affected and switches off the current, causing sudden catastrophic climate shift. The Earth is abruptly plunged into a new ice age.

Between Florida and the Bahamas the Gulf Stream has a maximum width of 80 km, a depth of 640 m, a surface temperature of 25°C and an average surface current 5 km/h.

Further north the stream steadily widens and is about 480 km wide off New York. South of the Grand Banks the stream joins the cold Labrador Current and, under the influence of south-westerly winds, moves north-east as the North Atlantic Drift at about 8 km a day.

The Gulf Stream moves huge quantities of water and heat from the tropics to northern latitudes and is vital in keeping Western Europe relatively mild and free of ice.

GPS and Satellite Navigation

The GPS (Global Positioning System) satellite navigation system was declared fully operational in July 1995. It is a US military system, but has been made available free for worldwide civil use. It now provides users anywhere on land, sea or air with highly accurate position, speed and time data continuously under all weather conditions. 24 satellites orbit the earth at an altitude of 20,200 km and, as the system depends absolutely on precise timekeeping, each satellite has four atomic clocks. GPS fixes are generally accurate to about 10–20m but the GPS receiver may need to be adjusted to the chart datum in use if different from WGS84 (World Geodetic System 1884). The proposed European Union system, Galileo, is behind schedule and is unlikely to be operational before 2012.

GPS is a pearl beyond price for navigators who previously would have had to rely on inertial systems, sextant fixes, radio direction finding and dead reckoning. However, the USA could switch off civil use at any time or satellites could be damaged, destroyed or jammed, so wise seafarers still keep the old-fashioned backup methods in reserve.

The most expensive films ever made (adjusted for inflation)

Cleopatra, 1963 ★ *Pirates of the Caribbean 2*, 2006
★ *Titanic*, 1997 ★ *Waterworld*, 1995

Three reasons why *Waterworld* cost $175 million

≈ The landing strip on the Hawaiian island of Kona was extended to allow a Boeing 747 to land and bring in all the equipment

≈ There was no filming for three days because of a hurricane warning. One morning was lost waiting for a storm tide to subside, and high winds constantly interrupted filming

≈ Kevin Costner's trimaran cost $190,000, and another $810,000 was spent on modifying it for the film. After it had taken three years to build them, and one day's filming had taken place, the director wanted all the catamarans sprayed a darker colour.

Titanic statistics

Flag of Registry	British
Date keel laid	31 March 1909
Launch date	31 May 1911
Build number	401
Length	269.06 metres
Beam	28.19 metres
Height	53.34 metres
Displacement	46,000 tons
Rivets Used	3,000,000
Crew	860
Passengers	2,500
Lifeboat capacity	1,178
Lifeboats	16 + 4 collapsibles
Propellers	3
Outer Props. Diameter	7.16 metres
Centre Prop. Diameter	5.03 metres
Watertight compartments	16
Funnels	4
Forward mast height	30.94 metres
Aft Main mast height	29.72 metres
Build cost	£1,500,000
Top speed	22.5 knots
Boilers	29
Engines	2 triple expansion + 1 turbine
Boiler pressure	215 p.s.i.
Wreck's depth	3,798 metres

Fourteen mistakes in the film *Titanic*

1. The fourth funnel on the real *Titanic* was added for aesthetic reasons, and only served as a ventilation shaft. In the film smoke is seen coming out of it.

2. When Jack and Fabrizio are seen waving as the *Titanic* puts out to sea, a piece of desert landscape can be seen in the background.

3. Captain Edward J. Smith served with the White Star Line for 32 years, not 26.

4. Third-class passengers could not see the first-class area.

5. The picture above the chimney in the smoking room shows New York harbour. But this picture hung in the *Titanic*'s sister ship, the *Olympic*.

6. Cal orders lamb with mint sauce for himself and Rose at lunchtime. But lamb was only on the menu for dinner.

7. At the time of the collision the *Titanic* was not steaming at full speed.

8. The species of dolphins shown in the film only live in the Pacific ocean.

9. In the film, Captain Smith is wearing contact lenses.

10. Jack refers to the Lake Wissota reservoir, not constructed until 1915, and to Santa Monica Pier, built in 1916.

11. The Statue of Liberty was not illuminated at that time.

12. The hymns sung at the church service on the day of the collision were not written until 1937.

13. The policeman's handgun was not available in 1912.

14. One of the Picasso pictures which supposedly goes down with the ship still exists today.

Lines of equality

~

Iso- means 'equal': all these terms denote a line on a
map linking points on the Earth

Isobath_____line of equal depth of water
Isogone_____line where magnetic declination is the same
Isohaline_____line connecting points of equal salinity
Isobathytherm ___line of depth of water with equal temperature
Isocheim___line of points with same average winter temperature
Isohel _____line of points having same duration of sunshine
Isochasm _____line of points where auroras are
observed with equal frequency
20° C-Isochryme ___line of mean winter temperatures: 20° is the
limit for coral reefs
Isothere _____line of mean summer temperatures.

Best beaches (USA)

~

(according to *USA Today*,
the country's best-selling daily paper)
≈ Ilio Point, Molokai, Hawaii
≈ Second Beach, Olympic National Park, Washington
≈ Cedar Tree Neck, Martha's Vineyard, Massachusetts
≈ Henley Cay, St John, US Virgin Islands
≈ St Joseph Peninsula State Park, Florida

Best beaches (France)

~

(as recommended by the travel guide, *Go France*)
St Jean-de-Luz, Pays Basque ≈ Villefranche-sur-Mer, Nice ≈
La Grande Motte, Montpellier ≈
Cap-Ferret, peninsula near Bordeaux ≈ Sète, Mont-Saint-Clair ≈
Argelès-sur-Mer, Côte Vermeille

Blue Flag Beaches bug parameters

Total coliform	Faecal coliform	Faecal streptococci	Bathing water quality
< 500	< 100	< 100	Good bathing water
500–10,000	100–2000	> 100	Allowed a few times during season.
> 10,000	> 2000		Possible sewage pollution. The blue flag must be removed.

The 3 types of bacteria are monitored at least every 2 weeks throughout the season.
All bug figures are per 100 ml of seawater.

Best beaches (UK)

1 Calgary Bay, Mull
2 Hell's Mouth, Gwynedd
3 Holkham, Norfolk
4 Holy Island, Northumberland
5 Lamorna Cove, Cornwall
6 Ringstead Bay, Dorset
7 Sinclair's Bay, Caithness
8 Westdale Bay, Pembrokeshire

The most exclusive seaside resorts in the world

(Source: on-line life-style magazine *AskMen*)
Fregate Island, Seychelles ≈ Te Mahia Bay, New Zealand
Round Hill, Jamaica ≈ Sabi Sabi, South Africa
Hideaway Island, Vanuatu ≈ Glover's Atoll, Belize
Upper Captiva Island, Florida ≈ Parrot Cay, Turks & Caicos
Lamu, Kenya ≈ East Winds Inn, St Lucia

Best aquariums

Georgia Aquarium, Atlanta ≈ 100,000 creatures of the deep, 500
species in 30 million litres of water
Kayukan, Osaka ≈ 35,000 creatures, 380 species in 5.4 million litres
Ocenario Lisbon ≈ 15,000 creatures, 450 species in 7 million litres
Zoo-Aquarium, Berlin ≈ 5000 fish, 450 species in 4 million litres
AQWA, Perth, Australia ≈ 400 species in 3 million litres
The Deep, Hull (England) ≈ 2000 creatures, 150 species in
2.5 million litres.

Hotels with a difference

Nyhed *Container* hotel,
    Refshaleoen, Denmark,
    107 euros.
    Email: jk@refshaleoen.dk

Crane Hotel,
    Harlingen,
    Netherlands (a converted dockside crane – sleeps two),
    299 euros with breakfast.
    Telephone: +31 900 540 00 01

Lighthouse Hotel Old Wicklow
  Head Lights, Ireland,
  An octagonal tower built in 1781,
  440 euros for five nights.
  Telephone: +353 1 670 47 33

Hotel *Ship* Amber Baltic,
  Miedzyzdroje,
  Poland,
  from 56 euros.
  Telephone: +48 91 3 22 85 00

Treehouse hotel Hackspecht,
  Vasteras, Sweden,
  Single room 13 m up in a tree
  from 76 euros, and
Underwater hotel
  Otter, Vasteras,
  Sweden,
  Floating, underwater hotel, where you sleep in an aquarium
  from 200 euros.
  Telephone for both: +46 21 83 00 23

Ice hotel
  Kakslauttanen,
  Finland,
  you can spend a night in an ice room on an ice bed
  from 200 euros.
  Telephone: +358 1666 7 100

The biggest fish markets

Tsukiji, Tokyo ≈ Tsukiji-Shijo underground station (Toei-Oedo-Linie)

Sydney Fish Market ≈ Bank Street Pyrmont, buses 501 und 443

Rungis, Paris ≈ Metro-Station Châtelet Les Halles, guided tours every second Friday of the month.

Fulton Fish Market ≈ Hunts Point, Bronx, New York, bus no. 6.

Underwater restaurants

Red Sea Star ≈ Eilat, Israel, telephone + 972 8 634 77 77

Ithaa ≈ Maldives, telephone + 960 66 8 06 29

Al Mahara ≈ Dubai, telephone + 971 4 301 77 77

Submarino ≈ Valencia, Spain, telephone + 34 961 975 565

Coming soon: *Ocean Park*, Hong Kong ≈ Poseidon, Bahamas ≈ *Nautilus* near Faro, Portugal.

'I feel seasick ...'

Children and older people tend to suffer less from kinetosis, or motion sickness; the sense of balance is underdeveloped in childhood, while among the over-50s it degenerates. Women, particularly during pregnancy or menstruation, more commonly suffer from motion sickness, as do Asians, and also people of a nervous disposition.

Advice:

1 Make sure you get enough *sleep* before you are due to travel.
2 On the day before your journey, avoid meat, mackerel, tuna, herring, and alcohol. They all influence your histamine level, which is what causes you to feel sick. Histamine is broken down by vitamin C, so suck or chew on a *Vitamin C* tablet before setting off.
3 If you still feel sick, try some *acupressure:* bend your arm in the direction of your elbow. Can you see the two prominent tendons

below the wrist? Situated between them, two thumb-widths below the wrist, is the 'neiguan point', which you should massage thoroughly.

4 *Before the journey*, eat a small meal, low in fat but rich in carbohydrate.

5 During the journey, look out of the window at the horizon, move your body – especially your head – as little as possible, and try to distract yourself.

6 You will feel the wave motion less on the lower decks of a ship. However, the air down below is not likely to be very fresh.

7 There are also *drugs:* the most popular are the *cholinolytica*; they contain atropine and/or scopolamine, and inhibit feelings of sickness. *Antihistamines* cause the least tiredness, while *neuroleptic* drugs have a powerful tranquillizing effect, and are therefore often given to children. But none of them are suitable for drivers, skippers or pilots. If in doubt, read the information on the packet, and consult your doctor or pharmacist.

8 And if you are going to be sick, do it *downwind*!

The Blue Riband

~

There has never been an actual 'blue riband' or ribbon, awarded for the fastest passenger-ship crossing of the Atlantic. It was not until 1935 that the British MP Harold Hales commissioned and awarded the Hales Blue Ribbon trophy, a three-and-a-half foot high silver, onyx and gilt statue, for the fastest passenger-ship crossing of the Atlantic. Nowadays the Hales Trophy is awarded by the US Maritime Museum in Long Island, New York. The classic route runs from Bishop Rock (the smallest island in the world, according to the *Guinness World Records*), the most westerly point of England, to the Ambrose Light in New York. The trip is usually quicker in the reverse direction, because of a following wind. Since Atlantic crossings vary in distance, the award goes on average speed.

Name of Ship · Year	Route	Duration of crossing	Average speed in knots
Sirius · 1838	Cork–Sandy Hook	18 days/14 hrs/22 mins.	(8.03)
Great Western · 1838	Avonmouth–New York	15 days/12 hrs/0 mins.	(8.66)
Great Western · 1839	Avonmouth–New York	14 days/16 hrs/0 mins.	(8.92)
Great Western · 1839	Avonmouth–New York	13 days/12 hrs/mins.	(9.52)
Columbia · 1841	Liverpool–Halifax	10 days/19 hrs/0 mins.	(9.78)
Great Western · 1843	Liverpool–Halifax	12 days/18 hrs/0 mins.	(10.03)
Cambria · 1845	Liverpool–Halifax	9 days/20 hrs/30 mins.	(10.71)
America · 1848	Liverpool–Halifax	9 days/0 hrs/16 mins.	(11.71)
Europa · 1848	Liverpool–Halifax	8 days/23 hrs/0 mins.	(11.79)
Asia · 1850	Liverpool–Halifax	8 days/14 hrs/50 mins.	(12.25)
Pacific · 1850	Liverpool–New York	10 days/4 hrs/45 mins.	(12.46)
Baltic · 1851	Liverpool–New York	9 days/19 hrs/26 mins.	(12.91)
Baltic · 1854	Liverpool–New York	9 days/16 hrs/52 mins.	(13.04)
Persia · 1856	Liverpool–Sandy Hcok.	9 days/16 hrs/16 mins.	(13.11)
Scotia · 1863	Queenstown–New York	8 days/3 hrs/0 mins.	(14.46)
Adriatic · 1872	Queenstown–Sandy Hook	7 days/23 hrs/17 mins.	(14.53)
Germanic · 1875	Queenstown–Sandy Hook	7 days/23 hrs/7 mins.	(14.65)
City of Berlin · 1875	Queenstown–Sandy Bank	7 days/18 hrs/2 mins.	(15.21)
Britannic · 1876	Queenstown–Sandy Hook	7 days/13 hrs/11 mins.	(15.43)

Germanic · 1877	Queenstown – Sandy Hook	7 days/11 hrs/37 mins.	(15.76)
Alaska · 1882	Queenstown – Sandy Hook	7 days/6 hrs/20 mins.	(16.07)
Alaska · 1882	Queenstown – Sandy Hook	7 days/4 hrs/12 mins.	(16.67)
Alaska · 1882	Queenstown – Sandy Hook	7 days/1 hr/58 mins.	(16.98)
Alaska · 1883	Queenstown – Sandy Hook	6 days/23 hrs/48 mins.	(17.05)
Oregon · 1884	Queenstown – Sandy Hook	6 days/10 hrs/10 mins.	(18.56)
Etruria · 1885	Queenstown – Sandy Hook	6 days/5 hrs/31 mins.	(18.73)
Umbria · 1887	Queenstown – Sandy Hook	6 days/4 hrs/12 mins.	(19.22)
Etruria · 1888	Queenstown – Sandy Hook	6 days/1 hr/55 mins.	(19.56)
City of Paris · 1889	Queenstown – Sandy Hook	5 days/23 hrs/7 mins.	(19.95)
City of Paris · 1889	Queenstown – Sandy Hook	5 days/19 hrs/18 mins.	(20.01)
Majestic · 1891	Queenstown – Sandy Hook	5 days/18 hrs/8 mins.	(20.10)
Teutonic · 1891	Queenstown – Sandy Hook	5 days/16 hrs/31 mins.	(20.35)
City of Paris · 1892	Queenstown – Sandy Hook	5 days/15 hrs/58 mins.	(20.48)
City of Paris · 1892	Queenstown – Sandy Hook	5 days/14 hrs/24 mins.	(20.70)
Campania · 1893	Queenstown – Sandy Hook	5 days/15 hrs/37 mins.	(21.12)
Campania · 1894	Queenstown – Sandy Hook	5 days/9 hrs/29 mins.	(21.44)
Lucania · 1894	Queenstown – Sandy Hook	5 days/8 hrs/38 mins.	(21.65)
Lucania · 1894	Queenstown – Sandy Hook	5 days/7 hrs/48 mins.	(21.75)
Lucania · 1894	Queenstown – Sandy Hook	5 days/7 hrs/23 mins.	(21.81)

Ship	Year	Route	Time	
Kaiser Wilhelm d. Große · 1898.		Needles – Sandy Hook	5 days/20 hrs/0 mins.	(22.29)
Deutschland · 1900		Cherbourg – Sandy Hook	5 days/15 hrs/46 mins.	(22.42)
Deutschland · 1900		Cherbourg – Sandy Hook	5 days/12 hrs/29 mins.	(23.02)
Deutschland · 1901.		Cherbourg – Sandy Hook	5 days/16 hrs/12 mins.	(23.06)
Kronprinz Wilhelm · 1902.		Cherbourg – Sandy Hook	5 days/11 hrs/57 mins.	(23.09)
Deutschland · 1903		Cherbourg – Sandy Hook	5 days/11 hrs/54 mins.	(23.15)
Lusitania · 1907		Queenstown – Sandy Hook	4 days/19 hrs/52 mins.	(23.99)
Lusitania · 1908		Queenstown – Sandy Hook	4 days/20 hrs/22 mins.	(24.83)
Lusitania · 1908		Queenstown – Sandy Hook	4 days/19 hrs/36 mins.	(25.01)
Lusitania · 1909		Queenstown – Ambrose	4 days/16 hrs/40 mins.	(25.65)
Mauretania · 1909.		Queenstown – Ambrose	4 days/10 hrs/51 mins.	(26.06)
Bremen · 1929		Cherbourg – Ambrose	4 days/17 hrs/42 mins.	(27.83)
Europa · 1930		Cherbourg – Ambrose	4 days/17 hrs/6 mins.	(27.91)
Europa · 1933		Cherbourg – Ambrose	4 days/16 hrs/48 mins.	(27.92)
Rex · 1933		Gibraltar – Ambrose	4 days/13 hrs/58 mins.	(28.92)
Normandie · 1935.		Bishop Rock – Ambrose	4 days/3 hrs/2 mins.	(29.98)
Queen Mary · 1936		Bishop Rock – Ambrose	4 days/0 hrs/27 mins.	(30.14)
Normandie · 1937		Bishop Rock – Ambrose	3 days/23 hrs/2 mins.	(30.58)
Queen Mary · 1938		Bishop Rock – Ambrose	3 days/21 hrs/48 mins.	(30.99)
United States · 1952		Bishop Rock – Ambrose	3 days/12 hrs/12 mins.	(34.51)

The Pale Blue Riband

The record set by the *United States* has been broken since 1989 by catamarans and speed boats. The New York Maritime Museum was forced by a court case to award the Blue Riband to catamaran ferries which had made only one Atlantic run, for advertising purposes.

Hoverspeed Great Britain, 1990 Ambrose–Bishop's Rock
    3 d/7 h/25 min. (36.97)
Catalonia, 1998 Ambrose–Tarifa (Spain) 3 d/9 h/55 min. (38.85)
Cat Link V, 1998 Ambrose–Bishop Rock 2 d/20 h/9 min. (41.28)

The records set by speed boats without passengers are recognized only by their owners:

Virgin Atlantic Challenger, 1986 Ambrose–Bishop Rock
    3 d/8 h/31 min.

Gentry Eagle, 1989 Ambrose–Bishop Rock 2 d/19 h/7 min.

Unusual units of measurement

Sverdrup – millions of cubic metres of water per second. Used to measure ocean currents.
Bale – cotton ready for shipping; American standard weight 21.22 kilograms, British 326.59 kilograms.
Smoz – 'sail-makers ounces', for measuring the weight of sailcloth: 42.828 grams per square metre.
TEU – (Twenty Foot Equivalent Unit): capacity of a standard container, corresponding to twelve register tonnes or 34 cubic metres.
Shackle (British), *shot* (American) – measure of length of anchor chain: 27.432 metres.

Tidal day – two changes of tide, i.e. the moon crosses the meridian twice, in 24 hours, 52.272 minutes.

Bind – old English unit of quantity for eels. One bind = 250 eels.

Canvas – the length of the (formerly canvas-covered) front of a racing boat.

Water column – pressure used to be measured with a water column whose top surface was sighted against an inch scale. Now only used in organ-building and the waterproofing of tents.

Mache – the radon content of mineral water is measured in Mache units.

Poise – physical unit for dynamic viscosity. Salt water, for example, has about five per cent more poise than fresh water, which is why it foams more.

Osmole – a measurement unit that defines the number of moles of a chemical compound that contribute to a solution's osmotic pressure. Blood plasma has 280–303 milli-osmoles per kilo; fresh water has 20 mOsm, sea water 1,010 mOsm.

Sydharb – unit of volume used for water in Australia. One sydharb – the amount of water in Sydney Harbour, which is approximately 500 gigalitres.

Midget Jellyfish threatens Queensland's Coast

The irukandji jellyfish which inhabits northern Australian waters packs a lethal sting in a miniature body of about 25 mm which makes it almost impossible to spot and avoid. This dangerous creature (*carukua barnesi*) was named after Dr Jack Barnes, who identified it in 1964. It has recently been found off Queensland's Fraser Island, hundreds of miles south of its usual hunting grounds and has halted filming of the Hollywood epic *Fool's Gold* starring Kate Hudson and Matthew McConaughey. These creatures are known to have killed two swimmers so far and an antidote has yet to be found.

Before the days of sonar or ASDIC echo-sounding, the lead line was the universal method of measuring the depth of water in fathoms (units of 6 ft/1.83 m).

The lead weight (about 3 kg) was heaved well ahead over the side of the moving ship and by the time the lead hit the bottom, the line would be approximately vertical. The linesman would look for a mark near sea level. If he saw one, he would call out, for instance, 'By the mark seven' but if no mark was visible at sea level an estimate was called out like 'Deep nine'. The bottom of the lead had a cup-shaped cavity filled with tallow which could indicate the seabed type, like sand or shell or, if clean, rock.

20-Fathom Lead Line Marks

1 fathom:	Deep one:	(no mark)
2 fathoms:	Mark two:	leather with two tails
3 fathoms:	Mark three:	leather with three tails
4 fathoms:	Deep four:	(no mark)
5 fathoms:	Mark five:	white cotton rag
6 fathoms:	Deep six:	(no mark)
7 fathoms:	Mark seven:	red wool rag
8 fathoms:	Deep eight:	(no mark)
9 fathoms:	Deep nine:	(no mark)
10 fathoms:	Mark ten:	leather strip with hole
11 fathoms:	Deep eleven:	(no mark)
12 fathoms:	Deep twelve:	(no mark)
13 fathoms:	Mark thirteen:	leather with three tails
14 fathoms:	Deep fourteen:	(no mark)
15 fathoms:	Mark fifteen:	white cotton rag
16 fathoms:	Deep sixteen:	(no mark)
17 fathoms:	Mark seventeen:	red wool rag

18 fathoms:	Deep eighteen:	(no mark)
19 fathoms:	Deep nineteen:	(no mark)
20 fathoms:	Mark twenty:	line with two knots

Ships called *Enterprise*

~

1705–1707	British 24-gun sloop
1709–1749	British 40-gun brigantine
1730–1764	British sloop, unarmed
1738– ?	British cutter, unarmed
1743–1748	British 8-gun sloop
1759– ?	British frigate, probably 12–40 guns
1774–1807	British 28-gun sloop
1775–1777	American battle sloop, 12 four-pounder guns and 10 swivel guns
1776 –1777	American privateer with 8 guns
1799–1823	American schooner with 12 six-pounders
1831–1844	American 10-gun schooner
1861	American balloon, unarmed
1864–1886	British battle sloop with four very large guns
1877–1909	American battle sloop with two large guns
1914–1918	American motorboat with one small gun
1919–1946	10-gun auxiliary
1930–1934	American racing yacht, unarmed
1935–1945	American trainer aircraft, unarmed
1936–1958	American aircraft carrier with 100 guns
1946–1959	American airship, unarmed
1959–1985	British patrol boat, probably handguns
1961–	American aircraft carrier with various anti-missile rockets
1976–	American space shuttle, unarmed
1976–1987	American racing yacht, presumably unarmed
1979–1991	Airship, unarmed

1981–	Patrol boat in Barbados, probably handguns
2151–	Starship* with ray-guns, phaser handguns
2245–2285	Starship, partly built in California
2286–	Starship, probably built on Earth
2293–	Starship from the constellation Antares
–2344	Starship built on Earth
2363–2371	Starship from Mars
2375–	Starship of unknown origin

* Spaceship *Enterprise* from the TV series *Star Trek: Enterprise*

Disaster scale

~

Epidemics ≈ Brockmann-Hufnagel-Geisel Scale

Meteor strikes ≈ Palermo Scale, Torino Scale

Tsunamis ≈ Murty and Soloviev Scales

Hurricane ≈ Saffir-Simpson and Fujita Scales

Earthquakes ≈ Richter-, Mercalli, and Shindo (JMA) Scales

Harsh conditions

Arctic pack ice measured in 'nordicity', the unit of measurement of northerliness.

Permanent pack ice	100
Pack ice in the Arctic Ocean	90
Nine-month pack ice	60
Six-month pack ice	40
Four-month pack ice	20
Less than one month pack ice	0

Whiteness

When light falls on an object or surface it is reflected to a greater or lesser extent. This can be expressed as an albedo value (from Latin 'albus' = white). The whole of the Earth and its atmosphere has an average albedo value of around 30%.

Fresh snow	90 %
Land surface	30 %
Sea surface	3 %
Sea foam	50 %

'Operation Habbakuk'

In October 1942 Geoffrey Pyke, a scientific adviser to Lord Louis Mountbatten and an eccentric inventor, proposed building a massive aircraft carrier made of ice, based in mid-Atlantic, to provide anti-U-boat cover for the vital shipping convoys.

Pyke enlisted Austrian glaciologist Max Perutz to help him develop pykrete, which was a mixture of ice and about 14% wood pulp or sawdust. This mix transformed the material into a tremendously strengthened super ice found to be, weight for weight, as strong as concrete. For maximum strength it was necessary to maintain a pykrete temperature of at least minus 15 °C by embedding refrigerated pipes in the structure.

HMS *Habbakuk*	
Length	609.5 m
Beam	91.4 m
Height	61 m
Hull thickness	12.2 m
Displacement	2,200,000 tons
Propulsion	26 electric motors of 1200 hp each
Speed	7 knots
Aircraft	150–200

Churchill attached great importance to the idea but declared that the scheme would only be possible if 'we let Nature do nearly all the work for us ... It will be destroyed if it involves the movement of very large numbers of men and a heavy tonnage of steel ... to the recesses of the Arctic night.'

The plan was eventually abandoned in January 1944 when US Navy engineers found that the amount of steel needed to freeze 2.2 million tons of pykrete in one winter in Canada was greater than the amount needed to build the entire ship of steel. By then long-range anti-submarine aircraft were available using centrimetric radar made possible by the British invention of the cavity magnetron. With Bletchley Park codebreakers having more consistent success with U-boat codes the threat to the Atlantic convoys was now under control.

Exxon Valdez oil spill disaster

The 300.5 m-long *Exxon Valdez* oil tanker ran aground on the Bligh Reef, Prince William Sound, Alaska, at 12.04 am on 24 March 1989. The tanker, which had strayed from the normal route, was fully loaded with 1,264,155 barrels of oil (over 53 million gallons). Of this cargo 11 million gallons/257,000 barrels were spilled.

Over 2,000 km of coast were oiled and around 300 km moderately or heavily contaminated. It was estimated that 250,000 animals died and most of the plankton in Prince William Sound were exterminated.

The clean-up took four summers. Exxon paid $2.1 billion to fund the up to 10,000 workers, 1000 boats and 100 aircraft. However, many believe that wave action from the winter storms did more to clean up the beaches than all the human effort.

The Exxon Shipping Company changed its name to 'Sea River Shipping Company'. The offending ship was repaired and renamed *Sea River Mediterranean*. It is now prohibited by law from returning to Prince William Sound.

Aquaholics 2006

These are the most popular names given to boats in the USA in 2006 according to the Boat Owners Association of the United States. *Aquaholic has consistently been in the top ten since 2002 (previous year's position in brackets):*

1 Aquaholic (2)   6 Knot Working (-)
  2 Second Wind (-)   7 Life is Good (-)
   3 Reel Time (-)   8 Plan B (-)
    4 Hakuna Matata (-)   9 Second Chance (-)
     5 Happy Hours (-)   10 Pura Vida (5)

How to change a boat's name

No
sailor,
even one without superstitious tendencies, would
change the name of a boat without some trepidation,
lest bad luck should follow. However, by taking
the following precautions all should be well
(although no guarantees are implied):

Remove *all* traces of the boat's present identity including name references on documents.

Write the old name of the boat in water soluble ink on a strip of metal.

Acquire two bottles of champagne and arrange for friends to attend the ceremony.

Stand on the bow of the boat and address Poseidon as follows:

'Oh great and glorious ruler of the deep, we humbly request that the boat's name on this metal plate be removed from your kingdom's records.' At this point, drop the piece of metal from the bow into the sea and pour overboard at least half of the champagne.

Now make a second appeal to the ruler of the deep:
'Oh mighty and omnipotent lord of the seas, we beseech that you record the name of this fine vessel henceforth and for ever as ...' At this point utter the new name and pour the second bottle of champagne into the sea (less a glass for each of the crew).

The new name should *not* be visible anywhere on the boat until completion of the ceremony. If the new name *has* already been painted on the transom, it should be covered until the end of the proceedings.

Windchill factor

~

The speed of the wind affects the air temperature, as the Antarctic scientists Paul Siple and Charles Passel noted when they were working for the US military during World War II.

Windspeed km/h Temp in C	0	10	20	30	40	50	60	70	80
20	20	19	16	15	14	13	13	13	13
15	15	13	10	8	7	6	5	5	5
10	10	8	4	1	-1	-2	-2	-3	-3
5	5	2	-3	-6	-8	-9	-10	-11	-11
0	0	-3	-9	-13	-15	-17	-18	-18	-19
-5	-5	-8	-15	-20	-22	-24	-25	-26	-27
-10	-10	-14	-22	-27	-30	-32	-33	-34	-34
-15	-15	-19	-28	-34	-37	-39	-41	-42	-42
-20	-20	-25	-35	-40	-44	-47	-48	-49	-50

Discovery Bay

~

Unsurprisingly, Discovery Bay has been a popular
place name. Discovery Bays can be found in:
Australia ≈ Barbados ≈
Dominican Republic ≈
Jamaica ≈ Hawaii ≈
Hong Kong ≈ California ≈
Canada ≈ Washington.

Coins with turtles on them

~

10

florin (Aruba) ≈
1 dollar (Bermuda) 500 cruzeiros
(Brazil) ≈ 1 cent (British Virgin Islands)
1 escudo (Cape Verde Islands) ≈ 10 cents
(Cayman) 5 tene (Cook Islands) ≈ 50 colones
(Costa Rica) 6 pence (Fiji) ≈ 500 sika (Ghana)
500 lire (Italy) ≈ 10 francs (Congo/Zaire) ≈ 1 ring-
git (Malaysia) 250 rufiyas (Maldives) ≈ 5 pcsos
(Mexico)1 dollar (Niue) ≈ 5 dollars (Palau) 100
balboas (Panama)≈ 25 pence (St. Helena) ≈ 4,000
kwacha (Zambia) ≈ 5 rupees (Seychelles) ≈ 24
shillings (Somalia) ≈ 3,500 dobras (São
Tomé & Principe) ≈ 1 seniti (Tonga) ≈
1 dollar (Tuvalu) ≈ 1 pound
(Cyprus)

The last survivors of the *Titanic*

~

Barbara Joyce West · Elizabeth Gladys 'Millvina' Dean

The percentage of women in the commercial fishing fleets

Belgium ≈ 3%	Ireland ≈ 0
Denmark ≈ 0	Italy ≈ 1%
Finland ≈ 0	Netherlands ≈ 5%
France ≈ 2%	Portugal ≈ 2%
Germany ≈ 0	Spain ≈ 1%
Greece ≈ 7%	Sweden ≈ 4%
Great Britain ≈ 1%	

Balsa trees

Light (0.12 grams per cubic centimetre) and very fast-growing (two metres a year) tree from South America, which reaches a height of eight metres free of knots. It was used by dwellers on the west coast to build boats for the coastal trade, and in Hawaii (before the introduction of synthetic resin) to make surfboards. Thor Heyerdahl sailed with his balsa-wood raft *Kon-Tiki* from Chile to the Polynesian island of Rairora.

Kelp

These giant blue-green algae are the fastest-growing organisms on the planet; they increase by as much as 30 centimetres a day. Like fungi, they are neither animal nor plant, but have a strange kingdom all to themselves; biologists refer to them as protistas or protoctists. In cold waters they are engaged in a perpetual struggle with sea urchins. If a storm, disease or hunger has driven out the sea urchins, the forests of kelp grow at a furious rate along the coast – until the sea urchins stage a come-back.

Kelp is rich in minerals: it contains potassium nitrate (saltpetre), used in the manufacture of gunpowder; alkali for soap; and sodium carbonate for glass. It is the carbohydrate algine in kelp that gives jelly, ice-cream and toothpaste their solid yet fluid consistency. Algine is also a versatile addition to textiles, giving them sheen, closeness of weave, smoothness, and fire- and water-resistant qualities.

Porsches v. fishes

Barracuda – acceleration 150 metres per second;
three times as fast as a Porsche
Pollock – brakes at 8.7 metres per second (in water!).
A Porsche brakes at up to 50 kilometres per second.
Atlantic tuna – achieves 90 km per hour over long distances.
Not bad for a fish that can weigh as much as a Porsche Carrera.
Butterfly fish – Has a turning circle of 6.50 centimetres, which corresponds to 50% of its body length.
A Porsche needs 12 metres.

Luminescent fish

Vampire squid ≈ Lantern fish ≈ Krill * Black dogfish ≈
Ctenophore or comb jelly ≈ Diaphanous hatchet fish ≈ Deep sea anglerfish ≈ Deep sea minnow ≈ Long-armed brittle star

27 April 1936: Joe Bowers climbed over a chain-link fence; shots were fired at him, and he fell 30 metres. He died of his injuries.

16 December 1937: Theodore Cole and Ralph Roe escaped. They were officially recorded as drowned.

23 May 1938: James Limerick, Jimmy Lucas and Rufus Franklin killed prison officer Royal C. Cline with a hammer and climbed on to the roof, where they tried to overpower another officer. But he shot Limerick, and the two others were given life sentences.

13 January 1939: Arthur 'Doc' Barker, Dale Stamphill, William Martin, Henry Young, and Rufus McCain were caught outside the walls on the western side of the island. Three gave themselves up, one was shot. Barker died.

21 May 1941: Joe Cretzer, Sam Shockley, Arnold Kyle and Lloyd Barkdoll took some guards hostage, but were overpowered, and given life sentences.

15 September 1941: John Bayless got as far as the sea, but gave up because the water was too cold. In the ensuing trial he tried to escape from the courtroom.

14 April 1943: James Boarman, Harold Brest, Floyd Hamilton and Fred Hunter took two hostages. They were discovered, and shots were fired at Boarman. He sank beneath the waves. Two days later the two others were caught – they had been hiding on the island.

7 August 1943: Huron 'Ted' Walters escaped from the prison laundry and was picked up on the island.

31 July 1945: item by item, John Giles gradually assembled a uniform in the laundry and calmly boarded an army boat. But his breakout was noticed immediately, and the boat was called back.

2 May 1946: Bernard Croy, Joe Cretzer, Marvin Hubbard, Sam Shockley, Miran Thompson, and Clarence Carnes took Officer

R. J. Miller's keys from him and opened the armoury. But as the keys did not fit all the doors, a fierce gun battle followed: 20 warders were wounded. The Marines were brought in to put an end to the mayhem. Coy, Cretzer and Hubbard died, Shockley and Thompson were sentenced to be executed in the gas chamber, and Carnes received a second life sentence.

23 *July 1956:* Floyd Wilson was caught after being at large on the island for twelve hours.

29 *September 1958:* Aaron Burgett and Clyde Johnson overpowered an officer on duty. Johnson was picked up out of the water, and Burgett drowned.

11 *June 1962:* John Anglin, Clarence Anglin, Frank Morris and Allen Clayton West managed to construct an electric drill and bored holes in their cell walls in order to get access to the utility corridor behind them. In the evening they put dummies in their beds and crept out through their escape hole. West was the only one who did not make it.

The others pushed their way out through vent holes, carrying home-made rafts and paddles. The next day a sailor noticed a corpse drifting by, but was unable to get a grip on it. Nothing more has been heard of the three escapees to this day.

12 *December 1962:* Carl Parker and John Paul Scott bent the bars of a kitchen window in the cellhouse basement, covered themselves in oil and wriggled out through the opening. They had made themselves a kind of airbed from gloves. Parker soon gave up, though, and swam back. Scott made it to the shore, but was so exhausted that he fell asleep on a rock underneath the Golden Gate Bridge. He was found by two boys, who thought he was a would-be suicide injured jumping off the bridge. They called the police. Scott is the only prisoner known to have swum right through the cold currents of San Francisco Bay. And just a week before Alcatraz was closed.

Halifax, Nova Scotia, has a deep, natural ice-free harbour and in 1917 was a garrison town as well as a naval dockyard. The inner harbour, known as Bedford Basin, was an ideal anchorage for assembling convoys and was used in both world wars.

On the morning of 6 December 1917 the Norwegian neutral ship *Imo*, in ballast, set off from the Basin bound for New York. At the same time, the French ammunition ship *Mont-Blanc* was entering the Basin and, after some ill-advised manoeuvres, the *Imo* struck the *Mont-Blanc* on the bow at the bottleneck known as the Narrows.

The *Mont-Blanc*'s cargo consisted of 2300 tons of picric acid used for making lyddite for artillery shells, 200 tons of TNT, 35 tons of benzol and 10 tons of gun cotton. The collision was minor but some of the drums of benzol broke loose, spilled on deck and caught fire. At this point the crew wisely decided to abandon ship. The ship burned for twenty minutes and gathered a large crowd of fascinated but ill-advised spectators.

At 9.04 am the *Mont-Blanc* exploded and the shredded remains were blasted in all directions. A mushroom-shaped cloud rose kilometres into the air and the 3000 tons of the pulverized ship rained down on Halifax. The ship's gun landed in Alboro Lake 2 km away.

Docks, ships, factories and churches were destroyed. 12000 houses were damaged, of which 1600 were totally destroyed and the death toll reached over 1900. That night the worst blizzard for years descended on Halifax.

After an official enquiry and a court case, an appeal to the Supreme

Court of Canada in May 1919 resulted in a final decision that both ships were equally at fault.

Each year at 9 am on the anniversary of the disaster a service of remembrance is held and bells ring out over the site of the largest man-made explosion before the atomic age.

Seas of the moon

Mare Anguis – Serpent Sea, ø 150 km
Mare Australe – Southern Sea, ø 603 km
Mare Cognitum – Sea that has become known, ø 376 km
Mare Crisium – Sea of Crises, ø 418 km
Mare Fecunditatis – Sea of Fecundity, ø 900 km
Mare Frigoris – Sea of Cold, ø 1596 km
Mare Humboldtianum – Humboldt's Sea
(German naturalist, 1769–1859), ø 273 km
Mare Humorum – Sea of Moisture, ø 389 km
Mare Imbrium – Sea of Showers, ø 1123 km
Mare Ingenii – Sea of Cleverness, ø 318 km
Mare Insularum – Sea of Islands, ø 513 km
Mare Marginis – Sea of the Edge, ø 420 km
Mare Moscoviense – Sea of Muscovy, ø 277 km
Mare Nectaris – Sea of Nectar, ø 333 km
Mare Nubium – Sea of Clouds, ø 715 km
Mare Orientale – Eastern Sea, ø 327 km
Mare Procellarum – Sea of Storms, ø 2,568 km
Mare Serenitatis – Sea of Serenity, ø 707 km
Mare Smythii – Smyth's Sea
(British astronomer, 1788–1865), ø 373 km
Mare Spumans – Foaming Sea, ø 139 km
Mare Tranquillitatis – Sea of Tranquillity, ø 873 km
Mare Undarum – Sea of Waves, ø 243 km
Mare Vaporum – Sea of Vapours, ø 245 km
(*There is also a Bay of Love, a Marsh of Sleep
and a Lake of Solitude.*)

The biggest oil platforms

Hibernia is sited off the coast of Newfoundland – 600,000 tonnes
below the waterline, 37,000 above. This drilling platform is the
only building in the world that could withstand the impact of
an iceberg. (The wreck of the *Titanic* lies nearby.)

Petronius stands in the Gulf of Mexico and is recognized as the
tallest free-standing building in the world, although only 75 of
its 610 metres can be seen above sea level.

Neft Dalshari is the first platform ever built, dating from 1947, and
has developed into a town in the Caspian Sea, with a population
of 5000 and a road network of 200 kilometres.

We all come from the sea

That is why our blood has the same level of salinity as the ancient
ocean at the point where our ancestors left it. If mammals at that
time had not had this salt content in their blood, they would have
become dehydrated through osmosis. Or they would have had to
drink massive amounts of seawater and shed the excess salt as
'tears', like the sea turtle. Another reminder of pre-historic times is
so-called 'surfer's ear', or exostosis. The ear grows a number of
lumps (exotoses) that grow into the ear canal and can almost close
the ear off, as though the sufferer were turning into a mermaid.
Cold-water surfers are particularly at risk; they are seven times
more likely to develop 'surfer's ear' than those in warmer waters.
They are advised to wear silicon plugs tailored to the ear, to prevent
water from getting trapped behind the blockage and causing a
painful infection.

Big ships that sank on their maiden voyage

~

Vasa – On 10.8.1628, the longest ship in the world at that time (69 metres) was launched in Stockholm. But two buoys soon marked the spot where her fatal maiden voyage ended: the ship keeled over, water rushed in through the gun ports, the ship capsized, and went down with all hands, of whom 50 died. The *Vasa* can be visited today in its own museum.

Maria Rickmers – One of only six five-master ships ever built, the ship was the first steam-assisted sailing craft and, at 3822 register tonnes, the biggest cargo ship of its day. But its very first voyage was ill fated. The Captain died in Singapore, and it was the First Officer who took the *Maria Rickmers*, carrying 57,000 sacks of rice, as far as Anjer Point in the Sunda Strait. It has never been seen since 24.7.1892.

Titanic – The biggest passenger ship in the world up to that time hit an iceberg off Newfoundland shortly before midnight on 14 April 1912, and sank after two and a half hours: 1523 people died. The position of the wreck is 41° 46' N, 50° 14' W, at a depth of four kilometres. A visit to the ship costs about £27,000: see www.deepoceanexpeditions.com

Bismarck – The first really powerful battleship acquired by the German Navy in the Second World War, it was one of the biggest and best-armed warships of its day. Its guns, firing 800-kilo armour-piercing shells, had a range of 36 kilometres; six planes could be catapulted from its deck; and armour plating accounted for 40% of its total displacement. When it went to sea in the spring of 1941, it was supposed to dominate the Atlantic. But after a week at sea and two battles (one successful, the other fatal), the *Bismarck* sank. Only 118 men survived from a crew of 2221. The ship lies at a depth of 4,800 metres at 48° 10' N and 16° 12' W. Deep Ocean Expeditions offers diving trips to this site, too.

The Lloyd's bell

Originally a French ship, *La Lutine* surrendered to the British after they won the Battle of Aboukir in 1793 and became HMS *Lutine*. When it sank six years later off the coast of Holland, it was carrying 1000 bars of gold and 500 bars of silver. One sixth of the treasure was recovered in the following decades. The ship's bell was also recovered: strangely enough, it was not the name of the ship that was engraved on it, but 'St Jean – 1779'. Since it was found, the bell has hung in the Lloyd's building in London. If there is bad news, the bell is rung once, and twice for good news. During the Second World War it only rang out on one occasion – when the *Bismarck* was sunk. Since then it has been rung every 11 November (Armistice Day), and for particular calamities such as the attack on the World Trade Center on 11 September 2001, or the Tsunami of 26 December 2004.

1953 East coast floods

In January 1953, the east coast of England was devastated by some of the worst flooding in living memory, exceeded only by the summer 2007 events. A number of extreme weather events combined to cause major flooding throughout Norfolk, Suffolk, Essex, Kent and the Thames Estuary. 307 people died, 100,000 hectares of eastern England were flooded, 24,000 homes were damaged or destroyed and over 30,000 people were evacuated. Damage was estimated at over £5 billion at 2006 prices. In the Netherlands 1800 people died and 200,000 hectares of polder country were flooded.

A fatal combination of factors caused this major storm surge, including a naturally high spring tide, extreme low air pressure and high winds from the NNW – the worst possible direction – which pushed a raised body or 'hump' of water south towards the bottleneck of the

Strait of Dover. Sea levels rose nearly 3 metres above mean high-water marks.

The predictions of net sea-level rise by 2100 by the UK Climate Impacts Programme (UKCIP) range from + 17 cm at best to + 77cm at worst. These estimates include isostatic subsidence of the land surface of – 0.8 mm per year.

In recent decades there has been a relative sea-level rise of about 2 mm per year.

Press-ganging
~

(*This entry is not suitable for those of a sensitive disposition*)
Kidnapping men for service at sea was called 'press-ganging' or 'shanghaiing'. They were made drunk and dragged on board ship. When they woke up the next day, they were far out at sea. A master of the art was the publican Joseph 'Shangai' Kelly in Portland. He pulled off his greatest coup when he found 22 men unconscious in a mortuary. They had broken in, and mistaken the embalming fluid for alcohol. Kelly loaded them on to barrows and took them on board a ship. The Captain assumed that their lifeless state and their low rattling noises resulted from a heavy night's drinking. But next morning he discovered the rattling had stopped: Kelly had sold him a crew of dead men.

Pirate attacks
~
2002 ≈ 370
2003 ≈ 445
2004 ≈ 329
2005 ≈ 276 (+ 440 abductions –
as many as in 1992, the first year for which figures exist)

Dangerous waters

Pirate attacks are most frequent in:
the Strait of Malacca and in Indonesian waters in general ≈
the South China Sea ≈ off the Nigerian and Somalian coasts ≈
Bab-el-Mandeb (the 'Gate of Tears' between Yemen and
Djibouti) ≈ off the east coast of South America

The weapons of choice for modern privateers

Kedge anchors ≈ Hand guns ≈ anti-tank weapons ≈
rocket-launchers

Inspiring islands

Cocos Island – This Pacific island belonging to Costa Rica is the
largest desert island in the world, and also one of the most
beautiful. Schools of hammerhead sharks, up to 500 strong,
teem in its waters. Among other hidden hoards, the legendary
Lima Treasure, now valued at more than £300 million, is buried
somewhere in this former pirate hideout. 'The existence of
hidden treasure on the island has been established', declared the
British Foreign Office in 1928. For 100 years adventurers hunted
for the gold in the rainforest, but to no avail. Robert Louis
Stevenson was the only one who made a fortune out of Cocos
Island; his novel *Treasure Island* was his greatest success.
What to do: diving; digging for treasure (illegal).
Warning: in case of diving accidents, there is no pressure chamber.
How to get there: only with organized diving groups; twelve days
for around £3000.

Isola di Monte Cristo – Near Elba, this island served as the setting
for the novel by Alexandre Dumas, *The Count of Monte Cristo*.

Though presenting a barren and forbidding appearance, the ravines in the granite rocks are home to some remarkable flora and fauna, including wild goats and a rare species of falcon. The only inhabitant is a very taciturn watchman, who lives in a little hut and keeps unwanted intruders out of this valuable biotope. Paradoxically, this idyllic spot is not far from civilization – yachts and the Italian coastline are always in view.

Islas Juan Fernandez – Because of his constant carping, in 1704 the sailor Alexander Selkirk from Lower Largo in Fife was cast away on a desert island in the Pacific, where he lived for nearly five years as a beachcomber. His adventures inspired Daniel Defoe to write *Robinson Crusoe*. This island of some forty square kilometres has a peculiar beauty, with its steep hills and deep canyons, and presents a challenge to hikers and mountain bikers. Seals sun themselves on the beaches, and in the amazingly warm water colourful fish abound, as well as the famous Juan Fernandez Lobster. The coast of Chile is 666 kilometres away.
What to do: spear fishing, walking.
Warning: no hospital, no bank.
How to get there: by air to Santiago de Chile (£600), and then fly with Air Taric (£280).

Ships that have disappeared in the Bermuda Triangle

General Gates, Hornet, Insurgent, Pickering, Wasp, Wildcat, Exprevier (1780–1824), *Rosalie* (1840), *Ellen Austin* (1881), *Timandra* (1917), *Cyclops* (1918), *Cotopaxi* (1925), *Suduffco* (1926), *Anglo-Australian* (1938), *Samkey* (1948), *Sandra* (1950), *Southern Districts* (1954), *Home Sweet Home* (1955), *Enchantress* (1958), *Ethel C* (1960), *Callista III* (1961), *Evangeline, Windfall* (1960–62), *Marine Sulphur Queen, Snoboy's* (1963), *Dancing Feathers* (1964), *El Gato* (1965), *Witchcraft* (1967), *Saba Bank* (1974), *Dutch Treat, Meridian* (1975),

L'Avenir (1977), *Hawarden Bridge* (1978), *High Flight* (1976), *Polymer III* (1980), *Sea Lure* (1983), *Real Fine* (1984), *Mae Doris* (1992), *Jamanic K* (1996), *Tropic Bird* (2000).

Titanic record 2007

The *Guinness World Records* has awarded 33-year-old Mark Colling from Llanelli, South Wales, a plaque for the world's largest matchstick model. He used 3.5 million matchsticks over 17 months to construct the 5.8 m, 1 ton, 1:100 scale model of the ship, complete
with iceberg.

Into the Med

On average the following spills, leaks or drains into the Mediterranean every year:

2 million tonnes of oil ≈ 800,000 tonnes of nitrogen ≈ 320,000 tonnes of phosphorus ≈ 60,000 tonnes of detergent ≈ 550 tonnes of pesticide

In and out of the Med

In:
+ 10,000 cubic metres of river water per second
+ 25,000 cubic metres precipitation per second
+ 1,100,000 million cubic metres of water from the Atlantic per second

And out:
– 110,000 cubic metres of evaporation per second
– 1,050,000 cubic metres outflow into the Atlantic per second

Recommended wrecks

A selection worth diving to:

SS *Yongala* (depth: 33 m, Townsville, Australia) · *Rainbow Warrior* (25 m, Bay of Islands, New Zealand) · SS *Thislegorm* (25 m, Shab Ali, Red Sea) · SS *Caldera*, SS *Escasell*, SS *Far Star*, SS *Ginger Screw*, SS *Glen View*, SS *Penelopez*, SS *San Andreas*, SS *Tropic* (3–20 m, all Banco Chinchorro, Mexico) · *Zenobia* (42 m, Larnaca, Cyprus) · *Bianca C* (25–55 m, Grenada) · *Endymion* (12 m, Salt Cay City, Turks & Caicos) · RMS *Rhone* (10–15 m, British Virgin Islands) · *Fujikawa Maru* (15–40 m, Truk Lagoon, Micronesia) · SS *Stavronikita* (6–33 m, Barbados) · USS *Saratoga* (15-70 m, Bikini Atoll) · *Sankisan Maru* (50 m, Truk Lagoon, Micronesia) · *Constellation* (10 m, Bermudas) · *Nord* (40 m, Tasmania) · *Superior Producer* (33 m, Curaçao) · *Antilla* (20 m, Aruba) · *Hilma Hooker* (33 m, Bonaire) · USS *Apogon* (50 m, Bikini Atoll).

The film *Deep Water* (2006) recounts the tragic voyage of Donald Crowhurst in the 1968/9 *Sunday Times* Golden Globe single-handed round-the-world yacht race. Nine sailors started but only one, Robin Knox-Johnston in *Suhaili*, finished.

Crowhurst's 12.2 m trimaran *Teignmouth Electron* was built under the pressure of the 31 October 1968 race start deadline and there was insufficient time to sort out snags. The most significant problem was leaking hatch seals which could not be fixed.

Once Crowhurst, burdened by financial difficulties, realized he could not complete the race, he hatched a plan to wait in the South Atlantic with apparent radio failure while pretending to be circling the globe. He would remain there until the other yachts came close, then make radio contact and sail for home to finish a close third behind Robin Knox-Johnston and Nigel Tetley.

First home would win the Golden Globe and the fastest finisher would win £5000 ($100,000 2006). By coming third and not receiving a prize, Crowhurst's fake logbook entries would not be scrutinized too carefully. This plan might have worked if Tetley's trimaran had not broken up on 21 May leaving him to be rescued from his life raft. Now Crowhurst would win the £5000 for fastest finisher and his logbooks would be meticulously checked and the fraud revealed.

On 1 July 1969, badly affected by loneliness and under enormous strain, Donald Crowhurst decided to end it all and abandoned his boat. It was discovered empty on 10 July by the Royal Mail Vessel *Picardy*.

Shakespeare and the sea

~

Doomed Clarence's dreadful sinking-to-the-sea-bed
dream from *Richard III* (I, iv, 24–31) could give the
doughtiest seafarers nightmares:

Methoughts I saw a thousand fearful wrecks,
A thousand men that fishes gnawed upon,
Wedges of gold, great anchors, heaps of pearl,
Inestimable stones, unvalued jewels,
All scattered in the bottom of the sea.
Some lay in dead men's skulls; and in the holes
Where eyes did once inhabit, there were crept –
As 'twere in scorn of eyes – reflecting gems ...

Blind solo sailors

~

Hank Dekker, 58 ≈ His attempted Atlantic crossing came to grief
after three days (though he did succeed in 1983 in sailing from
San Francisco to Hawaii single-handed).

Jim Dickson, 41 ≈ He was rescued from the water on the seventh
day of his attempted Atlantic crossing.

Geoff Hilton-Barber, 51 ≈ He successfully completed the 4350 sea
miles between Durban in South Africa and Fremantle in
Australia.

Youngest atlantic sailor

Michael Perham, aged 14, became the youngest sailor to sail single-handed across the Atlantic when he arrived at Nelson's Dockyard, Antigua, on 3 January 2007, after a six-week voyage from Gibraltar. His father, Peter, shadowed his son during the crossing in a similar yacht, a Tide 28.

Michael stopped off at Lanzarote in the Canary Isles and at the Cape Verde Islands during the 3500-mile voyage in his yacht named *Cheeky Monkey*.

Encounters with octopuses

Octopuses may be the most intelligent invertebrates, but the scattered remains of their prey give away their hiding places. So if you come across empty mussel shells, crab claws that have been sucked out, and the like, you should approach the octopus's cave with great caution, and avoid casting a shadow across the entrance if possible. The astute *cephalopod* will feel safe, and you will be able to observe it from just a few centimetres away. Note the changes in the colour of its skin. Its eyes are the best among submarine species. The pupils of a calm octopus are narrow, expanding when nervous. Curious by nature, it may stretch out a tentacle to feel you – but relax. If you happen to have a few shrimps on you (uncooked, of course) you can even feed it.

You are less likely to come across an octopus in open waters. If you do, do not attempt to catch it. You may damage its skin; parasites and deadly infection will be the outcome. Do not be disappointed if the octopus takes flight. If it moves closer to you out of curiosity, hold out your arm towards it with an upwards movement and open palm – it may perch on your arm. But do not panic if it explores you and your diving gear with its tentacles. Do not pull away, but wait until the octopus decides to move off.

Lord Nelson's pension

After the death of Horatio Nelson at the battle of Trafalgar on 21 October 1805, King George III gave his brother, the Reverend William Nelson, an earldom and Parliament granted him a generous pension. The pension amounted to £5000 a year (£3.7 million at present-day values) and a lump sum of £99,000 (£100 million today) to buy Trafalgar House, a stately home near Salisbury.

This property and annual pension granted to a hero's family by a grateful nation continued, with some reductions in the late nineteenth century, through the generations until the 'Trafalgar Act' of 1947. By this Act, Prime Minister Clement Attlee and the Labour government stopped the pension worth, by this time, about £400,000 a year today. The house was lost to swingeing death duties when the 5th earl died in 1951.

Coral reefs

The biggest and the best are in:
Great Barrier Reef · Belize · Florida Keys · Turks & Caicos

The 'Big Five'

Big game hunters have five items on their wish lists:
Lion, elephant, buffalo, leopard, and rhino.
Sport fishermen also have their own big five:
Barracuda, blue shark, swordfish, blue marlin and
    northern bluefin tuna.

Angling world records

(Source: *International Game Fish Association*)

Species	Weight in kg	Place	Date	Angler
Great barracuda	38.64	Christmas Island, Republic of Kiribati	11.4.1992	John W. Helfrich
Sea bass	205.93	Catalina Island, California	16.9.1980	Lillian Scott
Blue shark	239.49	Montauk Point, New York	9.8.2001	Joe Seidel
Blue marlin (Pacific)	430.92	Le Morne, Mauritius	16.12.1994	Maria Tomaini
Swordfish	536.15	Iquique, Chile	7.5.1953	Louis Marron
Hammerhead shark	449.51	Sarasota, Florida	30.5.1982	Allan Ogle
Marlin, Atlantic blue	636	Vitoria, Brazil	29.2.1992	Paulo Amorim
Northern bluefin tuna	678.58	Aulds Cove, Nova Scotia	26.10.1979	Ken Fraser

Spearfishing Records

(Source: *International Underwater Spearfishing Committee*)

Species	Weight in kg	Place	Date	Diver
Great barracuda	29.4	Japan	10.5.2005	Paul A. Smith
Blue shark	104.8	USA	9.8.1974	Robert Ballew
Short-tail stingray	220.4	New Zealand	12.5.1957	Ian Porter
Giant black sea bass	247.2	USA	1.9.1968	Bob Stanbery
Blue marlin	301.2	Brazil	7.1.2006	Carlos A. Sicupira
Atlantic bluefin tuna	296.6	Portugal	19.8.1997	Paulo A. da Costa Gaspa
Goliath grouper	364.7	USA	5.10.1949	Don Pindar

The 1979 Shark-tagging Championship

In the only Shark-Tagging Championship ever held off the Virgin Islands, staged in order to make a CBS television documentary, competitors had to leave a shark cage and touch as many sharks as possible in ten minutes, hitting them on the nose or catching them by the tail. There are very few sharks around the Virgin Islands, and the resulting CBS film, reduced to 20 minutes, cost a million dollars and is rated as one of the most boring events in television history. It has been broadcast four times all the same.

Parkinson's Law and the British Admiralty

C. Northcote Parkinson (1909–93) formulated the law which holds that all work expands to fill the time available for its completion, that administrators make work for each other and that managerial ranks swell inevitably at a rate of 5.7–6.56% annually.

Admiralty Statistics

Classification	1914	1928	Change
Capital ships in commission	62	20	- 67.74%
Officers & men in RN	146,000	100,000	- 31.50%
Dockyard workers	57,000	62,439	+ 9.54%
Dockyard officials & clerks	3,249	4,558	+ 40.28%
Admiralty officials	2,000	3,569	+ 78.45%

HMS Pinafore – Sir Joseph Porter's advice

Now landsmen all, whoever you may be,
If you want to rise to the top of the tree,
If your soul isn't fettered to an office stool,
Be careful to be guided by this golden rule –
Stick close to your desks and never go to sea,
And you all may be Rulers of the Queen's Navee!

Types of bait fish

Halfbeaks ≈ bigeye scad ≈ killifish ≈ golden shiners ≈ goggle eyes
≈ wobblers ≈ lake chubsuckers ≈ sculpins ≈ fathead minnows

Women who emerged from the sea

Aphrodite – The goddess of carnal love, 'foam-arisen Aphrodite',
was born from the sea-foam created near Paphos, Cyprus, after
Cronos castrated his father, Uranus, with a sickle, and threw the
testicles into the sea (Hesiod, *c.* 700 BC)
Venus – the Roman equivalent of Aphrodite was immortalized (so
to speak) by Sandro Botticelli in 1485.
Undine – Water spirit in *Liber de nymphis* by Paracelsus (*c.* 1530).
Arielle – The little mermaid in Hans Christian Andersen's fairytale.
Rheya – The dead beloved whom Kelvin meets again on the living
planet Solaris in Stanislaw Lem's novel of that name (1961).
Ursula Andress – in the James Bond film *Dr No* (1962).
Halle Berry – in the Bond film *Die Another Day* (2002).

Waterspirits ♀

~

Melusine (Celtic) ≈ Sirens (Greek mermaids)
Nereids (Greek sea nymphs)
Naiads (Greek freshwater nymphs)
Nixies (Scandinavian and Germanic)
Lorelei (German) ≈ Rusálka (Slavic)

Waterspirits ♂

~

Nokks (Germanic and Scandinavian mermen)

An A–Y of sea gods

~

Agir (Norse)
Beher (Ethiopian)
Czudo Morskoe (Slav)
Dylan (Welsh)
Ekke Nekkepenn (N. Fresian)
Fomoren (Irish)
Gefjon (Icelandic)
Jam (Canaanite)
Kanaloa (Hawaiian)
Llyr (Welsh)
Manannan (Celtic, *from which the
Isle of Man takes its name*)

Neptune (Roman)
Olokum (Yoruba)
Poseidon (Greek)
Ran (Germanic)
Sedna (Inuit)
Tangaro (Polynesian)
Veden Emae (Finnish)
Waruna (Indo-Aryan)
Yu-jiang (Chinese)

Sea monsters

Bahamut – (Arabic): a fish that lives in fathomless depths and carries the entire world. All the oceans of the world are contained in his nostrils.

Charybdis – (Greek): three times a day this monster swallows huge amounts of water in the Strait of Messina, and then belches it out again, causing whirlpools. Ships caught up in the maelstrom are doomed.

Leviathan – (Hebrew): a fire-breathing sea monster with which God plays in the afternoon, after he has studied the Tora, nourished the world, and sat in judgement upon it.

Scylla – (Greek): above the waist, she is a young woman, but her lower body is made up of hideous monsters, like dogs, who bark incessantly. She eats everything that comes near her.

Tiamat – (Babylonian): the primordial mother and goddess of salt water, married to Apsu, the god of fresh water. She is defeated in a battle with her great-grandson Marduk, he splits her in two, and from the two halves he creates heaven and earth.

Baywatch Babes

In order of popularity:
Pamela Anderson (1992–7) ≈ Gena Lee Nolin (1995–8)
Shawn Weatherly (1989–90) ≈ Nicole Eggert (1992–4)
Donna D'Errico (1996–8) ≈ Brande Roderick (2001–02)
Mitzi Kapture (1998–9) ≈ Erika Elenia (1989–92)
Brooke Burns (1998–2000) ≈ Simmone MacKinnon (1999–2000)

Steller's sea cow – vital statistics

The sea cow was discovered in 1741 by Georg Wilhelm Steller on Bering Island. Within 27 years it was extinct. We know little about *sirenia*, the scientific name of the order to which this 5-tonne animal belongs: all of the 20 specimens displayed in natural history museums are of uncertain provenance. Steller measured them very assiduously, however:

≈ Body length from the upper lip to the end of the right tail-fin, 741.5 centimetres

≈ Circumference around the neck, 204 centimetres

≈ Length of the nipples, 10.2 centimetres

≈ Distance between the tip of the upper lip to the sexual aperture, 490 centimetres

≈ Length of the sexual aperture, 26.0 centimetres

≈ Distance between the sexual aperture to the anal sphincter, 20.4 centimetres

Deep sea cables

3100 km, EURAFRICA ≈ St Hilaire (France) – Sesimbra (Portugal) – Funchal (Madeira) – Casablanca (Morocco)

6321 km, TAT-12/13 ≈ Green Hill (USA) – Land's End (England) –Penmarch (France) – Shirley (USA)

7104 km, CANTAT 3 ≈ Pennant Point (Canada) – Vestmannaeyjar

(Iceland) – Tjornuvik (Faroe Islands) – Redcar (England) – Blaabjerg (Denmark) – Sylt (Germany)

7552 km, PTAT-1 ≈ Manasquan (USA) – Devonshire (Bermuda) – Ballinspittle (Ireland) – Brean (England)

9500 km, SAT-2 ≈ Melkbosstrand (South Africa) – El Medano (Tenerife) – Funchal (Madeira)

11,428 km, SAFE ≈ Kapstadt (South Africa) – Durban (South

Africa) – St Paul (Réunion) – Baie Jacotet (Mauritius) –
Cochin (India) – Penang (Malaysia)

12,600 km, GEMINI ≈ Manasquan (USA) – Charlestown
(USA) – Oxwich (Wales) – Porthcurno (England)

13,000 km, ATLANTIS 2 ≈ Las Toninas (Argentina) – Rio de
Janeiro (Brazil) – Fortaleza (Brazil) – Dakar (Senegal) – Praia
(Cape Verde) – El Medano (Tenerife) – Funchal (Madeira) –
Concil (Spain) – Lisbon (Portugal)

13,000 km, SAT-3 ≈ Melkbosstrand (South Africa) – Luanda
(Angola) – Libreville (Gabon) – Douala (Cameroon) – Lagos
(Nigeria) – Cotonou (Benin) – Accra (Ghana) – Abidjan
(Côte d'Ivoire) – Dakar (Senegal) – Alta Vista (Canaries) –
Chipiona (Spain) – Sesimbra (Portugal)

14,000 km, AC1 ≈ Brookhaven (USA) – Land's End (England) –
Beverwijk (Netherlands) – Sylt – Brookhaven (USA)

15,428 km, TAT-14 ≈ Manasquan (USA) – Tuckerton (USA) –
Bude (England) – Katwijk (Netherlands) – Norden
(Germany) – St Valery en Caux (France) – Blaabjerg
(Denmark)

18,000 km, SEA-ME-WE 2 ≈ Jarkata (Indonesia) – Singapore –
Colombo (Sri Lanka) – Mumbai (India) – Djibouti – Aden
(Yemen) – Jeddah (Saudi Arabia) – Alexandria (Egypt)

21,000 km, PC1 ≈ Grover Beach (California) – Harbor Pointe
(Washington) – Ajigaura (Japan) – Shima (Japan)

24,593 km, TPC-5 ≈ San Luis Obispo (California) – Keawaula
Guam (Hawaii) – Miyazaki (Japan) – Ninomiya (Japan) –
Bandon (Oregon) – San Luis Obispo (California)

The mystery of the *Mary Celeste*

Near Gibraltar on 4 December 1872, the crew of the *Dei Gratia* came across an amazing sight: a ghost ship. Two sails were set, but there was not a soul on board.

When it was launched in 1861 the brigantine was called *Amazon*. The ship's owner died in mysterious circumstances, and on its maiden voyage the 30-metre ship was damaged when it ran into a fishing weir. While it was undergoing repairs, a fire broke out. On its second voyage the *Amazon* collided with a brig, which sank. For a few years it went about its business apparently without incident, until it was grounded twice in quick succession – at the same spot near Cape Breton. The repairs were too costly, and the ship was distrained. Her new owner called her *Mary Celeste*.

When the two-master was found drifting at sea, two hatches were open. The cargo hold was awash with water, and the seas were spilling in through a porthole. The crew of the *Dei Gratia* noted the following details: the galley was orderly, but the captain's cabin was in disarray, with sea-boots, clothes, a pipe and tobacco scattered about, and a keg of alcohol left open. The compass was damaged; the lifeboat was missing, as were the sextant, the chronometer and most of the ship's papers. The last entry in the log book dated from ten days earlier, 14 November.

The crew were never found, and the background to the story has never been explained. Sir Arthur Conan Doyle wrote a short story about the riddle, entitled 'J. Habakuk Jephson's Statement', part of the book *The Captain of the Polestar*. A seaquake, mutiny, and even the Bermuda Triangle have all been put forward as possible solutions to the puzzle.

The *Mary Celeste* was sold on a few more times, until finally it was deliberately run aground on the Rochelais Reef off Haiti. It was carrying rubber boots and cat food, but it was insured for a much more valuable cargo. The captain tried to float the ship off the reef and sink it in deeper water, but the agents of the insurance company arrived in time to see what was really in the hold. The last captain of the *Mary Celeste* was convicted of fraud.

Fish and shellfish farming

Worldwide in tonnes:

1950	638,577	1980	7,347,007	2000	56,687,909
1960	2,029,210	1990	16,827,906	2003	70,302,473
1970	3,525,872				

Oyster catch

Worldwide in tonnes:

1950	293,080	1980	300,149	2000	249,647
1960	204,955	1990	176,530	2003	199,517
1970	265,839				

Five sea battles that changed the world

1 *Salamis*, 480 BC –271 galleys and over 54,000 men on the Greek side. Over 500 warships and at least 100,000 men on the Persian side.

2 *Gravelines*, 1588 –The 130 ships of the Spanish Armada, with 26,000 guns and nearly 30,000 men, were opposed by 197 English smaller and lighter ships with 16,000 men. There were ultimately about 20, 000 dead from the Spanish ships after what remained of the Armada was scattered by a storm.

3 *Trafalgar*, 1805–27 ships of the line, four frigates, 2,048 guns,

23,400 men on the British side; 33 ships of the line, five frigates, 2,626 guns, 30,000 men on the French side. More than 5,000 dead.

4 *Jutland*, 1916 – The German High Seas Fleet consisted of 16 battleships, five battle cruisers, six ships of the line of the pre-Dreadnought generation, 11 smaller cruisers, 61 torpedo boats, 1194 guns; in the British Grand Fleet there were 28 battleships, 9 battle cruisers, 8 armoured cruisers, 26 smaller cruisers and 71 other ships, 1850 guns. 8645 dead.

5 *Guadalcanal*, 1942/43 – On the American side two battleships, five battle cruisers, 8 destroyers, various transport ships, and 29,000 soldiers. The Japanese had two battleships, four battle cruisers, 16 destroyers, various transport ships, and 30,000 soldiers. 30,000 killed.

Monitor v. Merrimac – the Battle of the Ironclads

The CSS *Virginia*, more commonly known as the *Merrimac*, was the first Confederate ironclad ship. It was built on the hull of the USS *Merrimac*, which had been scuttled and abandoned in April 1861.

On 8 March 1862 the newly converted ironclad steamed down the Elizabeth River into Hampton Roads to attack the blockading US fleet of wooden ships. *Merrimac* soon sank the 24-gun sloop *Cumberland* and followed up with the 50-gun frigate *Congress*. Just in time to save the rest of the Union fleet, the ironclad *Monitor*, designed by John Ericsson, arrived the next day and the two ships fought each other to a draw in which neither vessel was seriously damaged. A new era of naval warfare had begun.

The tallest lighthouses in the world

~

At 83
metres,
Isle de Vierge,
Brittany,
is the
tallest lighthouse
built in the traditional manner.
Yamashita Park, Yokohama, at 106 metres, has a
steel skeleton construction.

The Pharos of
Alexandria
is estimated
to have
been
between
120 and 140
metres high:
built 300–280 BC
It was
severely
damaged
by
earthquakes;
in 1303
and 1323,
and by
1480 it
had
disappeared
entirely.

Lighthouses where marriage ceremonies can be performed

USA:
Point Arena, California ≈ Cockspur Island, Georgia ≈
Cape Blanco, Oregon ≈ Montauk Point Lighthouse,
Long Island, New York

Great Britain:
Smeaton's Tower, Plymouth Hoe, Devon ≈
Rua Reidh, Scotland

Why lighthouse keepers go crazy

≈ When supplies ran out, they used to eat (tallow) candles.
≈ The heavy lenses and prisms floated in a mercury tank. Its fumes cost many keepers their sanity.
≈ Because two lighthouse keepers could enrage each other so much that only one survived – many shared their tower with a corpse for months – for a time in Britain it was stipulated that lights had to be manned by three keepers.
≈ The keeper, Henry Hall, was 94 when fire broke out in the Eddystone Light. In putting it out Henry unfortunately swallowed half a pound of lead, and was fatally poisoned.
≈ The wife of the keeper of Presque Isle Light, Michigan, was sent into such a frenzy by the solitude that he locked her in a cell beneath the platform. To this day she can still be heard shrieking.

Famous people with lighthouse keepers in the family

≈ The uncle of *Christopher Columbus* was a lighthouse keeper.
≈ Robert Stevenson, *Robert Louis Stevenson*'s father, was the most famous lighthouse builder in the 19th century.
≈ *Grace Darling*, the daughter of the keeper of Longstone Island

Lighthouse, England, became the first female media heroine when she rescued 9 shipwrecked people in 1838. A rose was later named after her.

Bishop Rock Lighthouse

~

The loss in 1703 of Sir Cloudesley Shovell's squadron of ships, including the *Association*, emphasized the need for an effective warning light for the Isles of Scilly. Work started on the Bishop Rock Lighthouse in 1847 and continued, against tremendous odds, until final completion in 1887.

Today it is one of the 72 lighthouses maintained by Trinity House which now use remote-controlled technology. Lighthouse keepers became a thing of the past in the 1990s and now the Central Planning Unit at Harwich, Essex, monitors and controls all operations.

Bishop Rock Lighthouse Facts

Position	49 52.3 N – 06 26.7 W
Engineers	James Walker & Sir James Douglass
Construction	5700 tons of granite
Tower height	49 metres
Light height	44 metres
Lamp	400 watt
Intensity	600,000 candela
Range	24 nautical miles
Light character	2 white group flashing every 15 secs
Fog signal range	4 nautical miles
Fog signal character	2 blasts every 90 seconds
Electrified	1973
Automated	1992

James Caird, saviour of Shackleton's expedition

The *James Caird* was the modified 8-m whaler in which Sir Ernest Shackleton (1874–1922) and five companions made their incredible open boat voyage from Elephant Island to South Georgia during the Antarctic winter of 1916.

After the sinking of the *Endurance*, the 28-man crew had left the ice pack in three small boats and managed against all odds to land on desolate Elephant Island on the tip of the Antarctic Peninsula.

In 16 days Shackleton and his hand-picked crew crossed 1300 km of the world's coldest and most treacherous seas. The *James Caird* was lashed by freezing storms and so laden with ice that the boat threatened to capsize before they managed to beach on the southern shores of South Georgia. From there, Shackleton and two of the strongest men still had to trek over 58 km of uncharted mountains and glaciers (with no mountaineering equipment) to reach the whaling station on the north side of the island.

Eventually, every crew member of the *Endurance* was rescued.

In 1994 Trevor Potts repeated Shackleton's journey. The boat itself can now be viewed in the North Cloister of Dulwich College, London.

**Polar bears that travelled
from Greenland to Iceland on ice floes**

1792 Two Polar bears were shot in Iceland. *1802*
Two Polar bears spotted. One was shot, and
nothing more was ever heard of the other one. *1874*
Six Polar bears landed in Hornstrandir and Mjóifjör
and were immediately dispatched. *1988* A Polar bear
landed in Iceland. It was sedated and immediately
flown back to Greenland. *1993* Sighting of a
swimming bear was reported.

Hunting and fishing in Greenland

*Greenland's hunters and fishermen are allowed to
catch the following each year:*
250 polar bears, 346 whales
100,000 seals

Whale catch

(1987–2002)

Species	Caught by indigenous people	Caught for 'Scientific Research'	Estimated population
Byrd's whale	–	190	60,000
Humpback whale	25	25	21,570
Fin whale	219	219	47,300
Gray whale	1685	1685	26,300
Bowhead whale	775		8000
Sperm whale	–	18	100
Sei whale	2	–	55
Minke whale	2146	6804	935,000

The North-West Passage

The appeal of being the first to find this fabled sea route linking the Atlantic with the Pacific north of Canada has been a lure to explorers for around 500 years. Amongst those who have tried to discover a reliable way through the ice were Sir Humphrey Gilbert, Martin Frobisher, John Davis, Sir Francis Drake, Henry Hudson, William Baffin, John Ross, William Parry, Sir John Franklin and Roald Amundsen.

Perhaps what all these explorers failed to find will finally become a reality through the ice-melting effects of global warming. A North-West Passage would save ships between 11,000 km and 19,000 km, depending on whether a ship can traverse the Panama Canal or has to go the long way round via Cape Horn.

Nanook of the North

This legendary 79-minute documentary about life in the Canadian Arctic shows kayaking, hunting, fishing, trading, igloo building and the daily struggle for survival of Nanook, an Inuit and his family.

In 1922 Robert J. Flaherty wrote and directed this first full-length documentary in the history of the cinema. The film, which required a 16-month expedition to Inukjuak, Hudson Bay, was sponsored by a French fur company, Revillon Frères.

Flaherty has been criticized for staging scenes and distorting reality, but this would have been inevitable in the hazardous locations used for filming.

Flaherty took with him around 23,000 m of 35 mm, blue-green

sensitive, orthochromatic film, 2 Akeley cameras and a printing machine. Film-washing was accomplished by keeping the nearest water hole (400 m away) free of ice over winter. A barrel of water was needed to wash every 10 m of film, amounting to 1,500 barrels over the course of the filming.

Swimming marathons

7 km from Robben Island, the island where Nelson Mandela was held prisoner, to Bloubergstrand ≈ 1 hr, 33 mins, 11 sec

7.5 km round South Africa's Cape Point, from Diaz Beach to Buffels Bay ≈ 2 hrs, 20 mins

8.5 km from Dassen Island to Yzerfontein, South Africa ≈ 2 hrs, 35 mins

11 km round Cape Agulhas, South Africa ≈ 2 hrs, 16 mins.

15 km across the Straits of Gibraltar ≈ 2 hrs, 51 mins

26 km across Cook Strait between New Zealand's North and South Islands ≈ 5 hrs, 4 mins

35 km – English Channel ≈ 7 hrs, 3 mins, 52 seconds

35 km across False Bay, South Africa ≈ 9 hrs, 56 mins

45.5 km around the island of Manhattan ≈ 5 hrs, 45 mins, 25 secs

70 km Manly Cove to Parramatta, Australia ≈ 15 hrs, 42 mins

110 km from the Spanish mainland near Jávea to San Antonio on Ibiza ≈ 25 hrs

204 km down Norway's longest fjord, the Sognefjord ≈ 21 days – (though this was not non-stop!)

Toughest races

Atlantic Rowing

In January from Gomera to Antigua in two- or four-man boats. 2550 sea miles. Record ≈ 39 days, 2 hrs, 35 mins, 47 seconds. Cost ≈ 22,350 euros per team, plus boat.

Vendée Globe

Solo round-the-world race approx 25 000 sea miles. Record ≈ 71 days, 14 hrs, 18 mins, 22 secs. Cost ≈ 3–4 million euros

Volvo Ocean Race

31 250 sea miles over nine stages around the world. Record ≈ 87 days, 10 hrs and 47 mins. Cost ≈ 10–20 million euros.

Sydney–Hobart Yacht Race

From Sydney (New South Wales) to Hobart (Tasmania) – 628 sea miles record ≈ 1 day, 40 mins and 10 secs. Cost ≈ 1998 six lives.

The Fastnet Race and Weston Martyr

Weston Martyr is credited with sparking the creation of the Fastnet Race by writing a letter to a yachting magazine in 1924 after competing in the Bermuda race aboard *Northern Light*. He wrote: 'It is without question the very finest sport a man can possibly engage in for to play this game at all it is necessary to possess, in the very highest degree, those hallmarks of a true sportsman: skill, courage and endurance.'

As a result of his enthusiasm the first Fastnet race started on 15 August 1925 with 7 yachts. They were to round the Isle of Wight, the Scillies and the Fastnet Rock, a total of 989.7 km.

The first winner was 56-foot pilot cutter *Jolie Brise* in 6 days, 14 hours, 45 minutes. This classic boat is still sailing today and won the 2000 Tall Ships Race overall.

The biggest ever Fastnet fleet of 303 yachts in 1979 was caught in a vicious storm that resulted in 17 deaths. Improved safety regulations were introduced thereafter and the legendary race has gone from strength to strength as a biennial event.

The Norfolk Broads and Weston Martyr

In the 1950s Weston Martyr, against his better judgement, was tempted by the prospect of unlimited Navy rum, pre-war gin and Suffolk cured ham to join two friends on a Broads cruise on the 35-foot yacht *Perfect Lady*.

Being an experienced ocean sailor he regarded the prospect of Broads sailing as 'pure hell' and something that would ruin his standing with his friends at the Royal Ocean Racing Club.

After many marine disasters and humiliations he had to admit that he had met his match. In a three-hour beat to windward along a 50-foot-wide river they tacked every nine seconds, or well over 1,000 times to progress 3 miles. 'If the Bermuda or Fastnet courses can provide any tougher job than that, I'm glad to have missed it,' he said. On his return to the Royal Ocean Racing Club he declared, 'I've just come back from the toughest cruise I ever made in my life. I've been sailing on the Broads!

'And now I'll be grateful if any of you men will ship me for the next ocean race, because after my week on the Broads, I need some soft, easy sailing and a nice quiet rest.'

The Ark

~

According to Genesis, it was 300 ells
long (about 133 metres), 50 ells wide
(22.30 metres), and 30 ells high
(13.40 metres). With these dimensions it would have had a
displacement roughly comparable to that of the *Titanic*. We
know that Noah was commanded to take on board a pair
of every animal, and seven pairs of the 'clean' animals
(domestic cattle) and birds. Attempts have been made to
take these figures literally; but it is impossible to do so.
Noah had 40,000 cubic metres at his disposal. Even if
he had left all the fish, invertebrates and insects on
land, he would hardly have had room to accommodate
even the original forms of each type of animal.
Instead of poodles, wolves and jackals, for
example, he would only have been able to take a
take a single representative of the dog
family on board.

Fish without swimming bladders

~

Moray ≈ sheatfish ≈ flatfish ≈ plaice ≈ sand eel ≈ shark ≈
sand tiger ≈ mackerel

Hunting starfish

~

Sea urchins can smell it, but they cannot escape the sunflower
starfish (*Pycnopodia helianthoides*). With a diameter of up to a
metre, it is the largest of the starfish, and with a speed of 3 metres
a minute it is also the fastest. It likes to grab the tasty echinoderm
with one of its arms (of which it has between 5 and 24) and suck
out the contents.

The sea as symbol

Standing for:

≈ the *mother* (according to the psychoanalyst Marie Bonaparte, referring to Edgar Allan Poe's stories)

≈ *God* (so say various literary critics about the sea in Stanislaw Lem's *Solaris*)

≈ *death* (in Shakespeare, Emily Dickinson, Thomas Mann, and many others)

≈ a *soul in turmoil* (the stormy marine paintings of the Romantics)

≈ the *unconscious* (interpretation of dreams)

≈ the *female sex* (in Goethe, Jules Verne and others)

≈ the *origins of evil* (Book of Revelation)

The settlement of the South Seas

Papua-New Guinea and Australia were settled from Asia 40,000 years ago. In the following 10,000 years settlers also went on to occupy the Solomon Islands and the Bismarck Archipelago. But how did they get from there to Vanuatu, Fiji, Tonga and Samoa?

The 'express' theory: 3,500 years ago, Lapita peoples from Taiwan and South-East Asia conquered the Pacific islands in the course of a few centuries.

The Bismarck Archipelago theory: according to this, the Polynesians are descended from the population of the Archipelago. The second wave of immigration contributed very little.

The rat theory: examining rats has brought a third version to the fore. The early Oceanians took the rodents with them in their canoes as a source of food. By analysing the rats' DNA, the settlement of the South Seas can be reconstructed as follows: the remote islands – Vanuatu, Fiji, Tonga and Samoa – were settled in a second wave of immigration directly from Asia (Taiwan); the Lapita people therefore bypassed Papua-New Guinea.

Elvis in swimming trunks

〜

Paradise, Hawaiian Style 1966
Clambake 1967
Easy Come, Easy Go 1967

The origins of American place names

〜

Barbados ('bearded') – Portuguese sailors were struck by the beard-like roots of a Caribbean fig-tree.

Canada – The Native American Iroquois, giving directions to Jacques Cartier in 1535, indicated: 'That's the way to the *kanata*' (a large village). (An alternative version has it that a disappointed Spaniard exclaimed 'Ca nada – There's nothing here!')

Panama – Meaning 'many fish', this was the name given by the natives to a village near the present capital.

Chile – 'End of the world' in the native language, Aymara? Or does the name derive from 'chi', a bird with yellow spots on its wings? Or is it to do with 'tschili', meaning 'cold' in another native language?

Brazil –Supposedly derived from the Portuguese discovery of a New World tree related to the Old World brasil tree. An alternative explanation relates the name to the mythical island of Hy-Brasil, 'somewhere to the west of Ireland', which many seafarers tried to find, including John Cabot. (See the entry on 'Phantom Islands', above.)

Turks & Caicos – On the old sea charts sailors were warned about 'Turks', a common expression at the time for pirates. Caicos comes from *Cay hico* (Carib for chain of islands) or from *Casco* (Spanish for ship's hull). But there is also an eccentric Turkish high school teacher who is convinced that the name proves the Turks discovered America 25 years before Columbus.

Nautical expressions II

Lazy Jacks – allow the mainsail on a yacht to be easily reefed and stowed by restraining the sail between lines running from the mast near the 'hounds' (= attachment point for the shrouds up the mast), to the boom.

The bitter end – there used to be a frame of oak timber on deck called a bitt, consisting of two oak bitt-pins and a cross-piece, to which anchor cables or ropes could be attached. A 'bitter' is one turn of the line around the bitt. 'To pay a rope out to the bitter end' thus means to let the anchor out until it strikes the seabed.

Groggy – the British Admiral Vernon always wore a coat made of grogram (or grosgrain), a rough blend of wool and silk, and this earned him the nickname Old Grogram. He gave orders that the daily rum ration should be diluted with water; the sailors called this mixture 'grog', and anyone who drank too much of it became groggy.

To let the cat out of the bag the cat o' nine tails was a lash carried by the Mate in a bag.

Mayday – the SOS call, derived from the French 'M'aidez', 'help me!'

Bluetooth – the wireless network takes its name from Harald Bluetooth (910–77), a warlike Viking who united Scandinavia. The Swedish firm Ericsson called their new technology after him, but it is not clear whether Harald really had a blue tooth, or whether 'Blåtand', as he is called in Swedish, is a reference to his (for a Viking) dark skin and hair colour.

Place names with Christian associations

The *Easter Islands* belong to Chile. There are two sets of *Christmas Islands*, one now bearing the name of Kiribati (pronounced 'Kiribash'), and the other a small atoll off the coast of Australia. *Advent Island* in the Antarctic was given its name on a December day in 1956, and *Conception Island*, named for Mary, is not far away. There are New Year Islands in New Zealand and off California (Año Nuevo Island). *Ascension Island*, St Helena, and Tristan da Cunha form a British territory in the South Atlantic, and *Assumption Island* is part of the Seychelles.

Music of the deep

North of Hawaii there is a range of seamounts (mountains with summits below sea level), which are named after composers. Going from 165° West to 157.3° West they are:

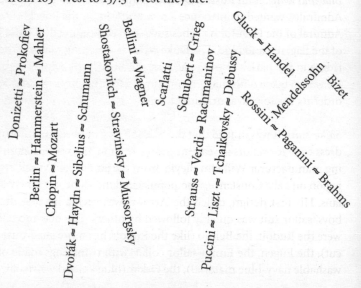

Biscuits of the deep

~

(Peake Frean and Huntley & Palmer are old biscuit companies.
Their deeps and ridges lie east of the Azores.
Right at the end is Crumb Seamount.)
Peake Deep · Frean Deep
Huntley Ridge · Palmer Ridge

Underwater Tolkien

~

Rohan Seamount ≈ 54° 45' N – 22° 20' W
Gondor Seamount ≈ 54° 15' N – 23° 50' W
Eriador Seamount ≈ 54° 50' N – 25° 20' W

Maritime fashion

~

Blue and white: John Russell, Duke of Bedford and First Lord of the Admiralty, wanted to introduce a new uniform for the Royal Navy. Admiral of the Fleet John Forbes suggested combining the colours of the flag, red and blue. The Duke replied: 'The King saw my wife riding in the park a few days ago, in a blue suit with white lapels. He was so taken with this combination that he has been pleased to order its use for naval uniforms.'

Sailor suits: It was in 1846 that the British Navy imposed a uniform dress code for sailors at sea. But strangely enough it was the clothing manufacturer Wilhelm Bleyle, from the landlocked Voralberg region on Lake Constance, who popularized the sailor suit for civilians. His first design, called the 'Arthur', was such a hit that this boys' sailor suit was quickly followed by others. The new models were the Rudolf, the Benno (like the Rudolf, but more generously cut), the Eugen, the Emil ('sailor collar' with trimmings made of washable navy-blue material), the Oskar (blue knitted waistcoat), and the Harold (with a hand-knitted, colour-fast anchor). For girls there were blouses named Louise, Else and Lotte. It was very lucky for the firm of Bleyle that BASF (the big German chemical conglomerate) had just brought on to the market a synthetic blue dye to replace the traditional one derived from the expensive raw material, indigo. Blue-and-white-striped shirts (seersucker) were also popular in Britain and America at the time, but nowhere was the craze for sailor suits as widespread as in Germany. Nearly every child had one.

Double-breasted jackets: It was sailors in the late 19th century who were the first to wear jackets with two rows of buttons, and on dry land the style only caught on at the beginning of the 20th century.

Three-quarter-length trousers: In the 1950s Emilio Pucci di Barsento, inventor of tapered trousers, was inspired by the fishermen of Capri to create his so-called 'Capri trousers'. This three-quarter-length look was copied from the fishermen's habit of rolling up their trouser legs when hauling their boats ashore.

Turtle/roll neck: first appeared around 1890 as sportswear, for footballers or riders, and was adopted by workmen and seamen at the turn of the century. The navy adopted it into their uniform code, and early feminists wore it as a 'unisex' symbol.

Hook & eye: In the 17th century, some seaman had a brilliant idea. He replaced the waistband, which expanded when wet, with a hook and eye, still in use for necklaces and brassieres.

Sailor collar: A rectangular collar, fashionable sea-going wear from 1839 onwards, proved easier to sew than its competitor the triangular collar, which was also popular.

Jersey/sweater/pullover: Tightly knitted outerwear was probably introduced by fishermen on Jersey. Wool was formerly the only material that would keep you warm even when wet. Ordinary fishermen and sailors were only allowed to wear striped jerseys; plain-coloured ones were reserved for officers. Around 1890 the woollen garment was taken up by sportsmen who wanted to work up a good sweat; at any rate, this is the origin of the American term 'sweater'. When the jersey became everyday wear, it was initially called a 'jumper' in English, and from about 1930 a 'pullover'. In the 1940s knitwear could be made increasingly tight-fitting, emphasizing the charms of the 'sweater girls' – Lauren Bacall, Lana Turner and Jane Russell.

Bell bottoms: Introduced in 1817, flared trousers could be rolled up to the knee when a sailor was swabbing the deck, or had to work in flooded areas. They could also be removed quickly if a sailor fell overboard or was forced to abandon ship.

When did the Vikings start wearing horns?

'The helmets of the Cimbri [Germanic tribes] are made to resemble the heads of wild beasts', wrote Plutarch. Another historian, Diodorus Siculus, reports seeing antlers and horns on the helmets of the (Celtic) Gauls, but the Thracians were also known for this kind of decorated headgear. Such helmets were far too impractical to wear in battle, however; they were only used for ceremonies and festivals. Two bronze-age horned helmets have been found, in Denmark and at the bottom of the Thames, but the horns are more suggestive of ice cream cones than oxen. From 1616 onwards painters liked to depict ancient Teutons wearing horned helmets, and this was a feature taken up by the Romantic painters, who were not too concerned whether the wearers were Germanic or Celtic. It was in 1820, art historians agree, that the Swedish painter Gustav

Malmström started to decorate the helmets of Vikings with classical horns. The fashion was given added impetus by the amateur archaeologist Axel Holmberg, who claimed to have seen bulls' horns on Viking rock drawings. The big breakthrough came with Wagner's opera *The Ring of the Nibelungs*. The Valkyries wore cows' horns stuck on to their helmets. From there they spread inexorably to children's stories and comics.

Why pirates wear earrings

An
acupuncture point
in the ear-lobe improves
vision. ≈ An earring helps prevent seasickness. ≈ Gold worn in
the ear is a practical investment. ≈ If
a sailor should be washed up dead
on shore, the ring will cover the
funeral expenses. ≈ A ring is a
souvenir of a particular
voyage. ≈ It was the
fashion.

The Oak Island money pit mystery

For over 200 years hopeful treasure hunters (including the 27-year-old Franklin D. Roosevelt in 1909) have explored a mysterious and tantalizing pit on Oak Island off the coast of Nova Scotia.

The lure of pirate gold has tempted many to spend their own money (over $1 million has been sunk in the pit to date) on major digging expeditions. It has also caused the death of six diggers. The 4 m diameter shaft has been excavated by hand to a depth of 28.6 and drilled further to 53 m. Findings include flagstones, oak logs, charcoal, putty, coconut fibre, cement, an inscribed stone, spruce planks and a variety of metal fragments. The pit was protected by a booby-trap flood tunnel at 33.5 m, which has caused further difficulties for the treasure seekers. The root cause of all this, an inscribed stone, which has since vanished, was deciphered in 1866 by a Halifax University professor: 'Forty feet below two million pounds are buried.'

The current searcher is Daniel Blankenship and his Triton Alliance group which continues the longest-running treasure hunt in the world.

Principality for sale

Probably the world's smallest self-declared country, 11 km off the coast of Harwich, Essex, is for sale (as of January 2007) for £65 million. The 550-square-metre ex-naval defence steel platform supported by two concrete towers was built in 1941 and was originally known as Roughs Tower. Sealand has been occupied by 'Prince' Roy Bates since 1967. Access to his principality is by helicopter and boat. (See 'States located on Artificial Islands', above)

Seamen's superstitions

Women on board bring bad luck, but if they ever *are* on board, they must wear a tam-o'-shanter. Women's hair is extremely dangerous.

Sunday is a good day to put to sea. But *Friday* will bring trouble and danger.

If you are *becalmed*, scratch the mast with an old nail to raise the wind. Alternatively, stick your knife in the mast on the bearing the wind is desired to come from. Or knot a short length of rope; a single knot for a light breeze, two for fresh breezes, and three for strong winds. Whistling for a wind is also effective, but *you must not whistle on board ship* except when becalmed, or you might offend the wind spirits and bring on a gale.

If you have had *to tack* for a long time through the wind and you encounter another ship, you should throw a broom towards the oncoming ship, and then the wind will back round. You will then have the wind astern, and the other ship will have to do the hard work of tacking.

The *soul* of anyone who dies on board will enter into a bird, and it will only show itself to presage another *death*.

If you have a pig and a cockerel *tattooed* on your feet, you must not go overboard. Neither of these animals can swim.

If you light your *cigarette* from a candle, a seaman will die. (Many seamen earned a living from making matches during the winter.)

When a ship is *named*, an earlier name can only be repeated for a later vessel if the predecessor went out of service honourably, through being sold to another owner, scrapped, or lost by enemy action. The name of a ship destroyed by fire, or lost in a collision or grounding, should not be repeated.

Long-distance rowers

~

John Fairfax rowed single-handed from the Canaries to Florida in
    180 days in 1969 = 8595 km
Peter Bird rowed single-handed from San Fransisco to the Great
    Barrier Reef in 294 days in 1982/3 = 13979 km

The Royal National Lifeboat Institution

Since the RNLI was founded in 1824, its lifeboats have saved more than 137,000 lives. In 2004 there were:

> 7,656 launches – an average of 21 a day
> 433 lives saved – an average of more than 1 a day
> 7,507 people rescued – an average of 21 a day

The RNLI is a registered charity independent from Government and continues to rely on voluntary contributions and legacies for its income. In 2004 it cost approximately £119 million (around £325,000 a day) to run.

RNLI Hovercraft

Hovercraft were introduced into the fleet in 2002 and can operate in mud, sand and very shallow water. They are particularly useful for shoreline searches. There are 4 in operation at the moment at Hunstanton, Norfolk; Morecombe, Lancashire; New Brighton, Merseyside and Southend on Sea, Essex.

Length	8 m
Weight	2.4 tonnes
Range	3 hours at max. speed
Speed	30 knots
Crew	2–4
Construction	Marine aluminium + composites
Launch type	Bespoke trailer

Time bomb in the Thames

In August 1944 the SS *Richard Montgomery*, a 7146-ton US Liberty Ship carrying 6000 tons of munitions, crossed the Atlantic and arrived in the Thames Estuary en route to her final destination of Cherbourg in support of the D-Day landings. During the night of the 20 August 1944 she ran aground on the Sheerness middle sand.

Despite desperate attempts to lighten the ship, the forward end began to flood and she broke into two pieces and sank. Only about half of her cargo was salvaged. That cargo of munitions for the US Air Force included:

    13,064 general purpose 250 lb bombs
    9022 cases of fragmenting bombs
    7739 semi-armour piercing bombs
    1522 cases of fuses
    1429 cases of phosphorous bombs
    1427 cases of 100 lb demolition bombs
    817 cases of small arms ammunition

Over sixty years later the wreck remains on the sandbank with her masts clearly visible at all states of the tide. There are approximately 1400 tons of explosive left on board in the submerged holds.

Should it explode it could be the world's biggest non-nuclear explosion, more destructive even than the 1917 ammunition ship explosion in Halifax, Nova Scotia.

Sheerness would cease to exist; every building would be flattened by the explosion and the resulting tidal waves.

Two Liberty ships are still afloat: the SS John W. Brown *and the SS* Jeremiah O'Brien. *Both are museum ships which still put out to sea regularly.*

Morse Code

A	•—	K	—•—	U	••—	1	•————
B	—•••	L	•—••	V	•••—	2	••———
C	—•—•	M	——	W	•——	3	•••——
D	—••	N	—•	X	—••—	4	••••—
E	•	O	———	Y	—•——	5	•••••
F	••—•	P	•——•	Z	——••	6	—••••
G	——•	Q	——•—			7	——•••
H	••••	R	•—•			8	———••
I	••	S	•••			9	————•
J	•———	T	—			0	—————

Full stop	•—•—•—	Semi-colon	• • •
Comma	——••——	Hyphen	—••••—
Colon	———•••	Plus	•—•—•
Question Mark	••——••	@	•——•—•
Exclamation mark	—•—•——		

SOS

When Morse code was first used, the distress call was 'CQD', which could be remembered by the phrase 'Come Quick, Danger!' But the Morse sequence 'dash-dot-dash-dot-dash-dash-dot-dash-dash-dot-dot' was too long-winded, which is why at the Berlin Radio

Conference in 1906 it was agreed that 'SOS' should replace it. Three short, three long, three short. The mnemonic for this was 'Save Our Souls'. The wireless operators on the *Titanic* initially sent out the familiar 'CQD', but eventually switched to the new 'SOS'.

The last Morse message from Australia

On 1 February 1999 the new Global Maritime Distress Safety System (GMDSS) was introduced. Morse has not been used since. 'THIS IS THE FINAL MORSE TRANSMISSION FROM THE TELSTRA MARITIME COMMUNICATIONS NETWORK.WE CONCLUDE OUR FINAL CW WATCH AFTER 87 YEARS OF CONTINUOUS SERVICE WITH PRIDE AND SADNESS. TELSTRA, THE AUSTRALIAN MARITIME SAFETY AUTHORITY AND THE BUREAU OF METEOROLOGY WISH ALL SEAFARERS FAIR WIND AND FOLLOWING SEAS. MARCONI IF YOU CAN HEAR THIS WE SALUTE YOU 73S – 31ST JANUARY 1999 2359 UTC + SK'

Sea music I

Sea shanties

A sea shanty or chanty (probably from French 'chanter') was used to help a team of sailors work safely and in rhythm with the work being done on merchant sailing ships. On board navy ships shanties were rarely used. There were also sea songs, unrelated to work, which helped entertain the crew when off watch. By the 1880s sailing ships were increasingly being replaced by steamers and the use of shanties faded away.

Different types of shanty suited different shipboard tasks like capstan or windlass turning, pumping and short and long drag halyard hauling.

Capstan turning needed a sustained, steady rhythm for raising or lowering the anchor, so a repetitive but very catchy shanty like 'The Drunken Sailor' was ideal.

> What shall we do with the drunken sailor?
> What shall we do with the drunken sailor?
> What shall we do with the drunken sailor?
> Earlye in the morning.
> Way, hay up she rises,
> Way, hay up she rises,
> Way, hay up she rises,
> Earlye in the morning!

A short-haul halyard shanty had to reflect quick hauls over a short time period, like 'Boney was a Warrrior'

> Boney was a warrior
> Oh aye, oh
> Boney was a warrior
> John Franzo. (From Jean François)

> Boney marched to Moscow
> Oh aye, oh
> Boney marched to Moscow
> John Franzo.

Pumping ship seemed like a never-ending tedious task but a suitable shanty could relieve the drudgery, like 'Fire down Below'.

> Fire in the cabin, fire in the hold,
> Fire in the strong-room melting the gold,
> Fire, fire, fire down below
> Fetch a bucket of water
> Fire down below.

Now the 'International Shanty and Sea Song Association' exists to encourage interest in this evocative relic of the days of sail. To listen to the shanty tunes try www.contemplator.com/sea.

Sea music II

Fado

Whether they lie with the Moors' lament over their defeat by the Christians, or Brazilian slaves bewailing their fate in a minor key, the origins of these songs of lament are obscure. Both of these sources can be heard in the music: it contains Latin American elements like *Lundum* or *Mondinha*, alongside musical arabesques. Two guitars are played, often accompanied by a *viola baixa*. The popularity of Fado (literally 'fate') spread from the dockside bordellos of Lisbon to become the national music of Portugal. The lyrics are about missing seafarers whose return is awaited with longing. But the sea does not willingly yield up its own. Other themes are, of course, lost love, the blows of fate, and a general mood of melancholy.

Fado stars, old and new
Maria Severa (around 1860) ≈ Artur Paredes (around 1930)
Amália Rodrigues (around 1950) ≈ Carlos Zel (around 1970)
Madredeus (Teresa Salgueiro) (around 1990) ≈ Misia (today)
Nelly Furtado (today) ≈ Mariza (today)

Calling all sponges

In minute electronic impulses, glass fibre cables carry an enormous amount of information, such as telephone connections, the Internet, television pictures. They are highly sensitive, break easily, and must be manufactured at high temperatures; a typical high-tech product of the digital age. But on the seabed, for millions of years, an extremely primitive creature has been doing it better. *Euplectella*, a variety of sponge, holds on to the seabed with nails that convey light just as well as a glass fibre cable, but can also be knotted, and will grow even at low temperatures.

Eight ways the sea could save the planet

Power stations exploiting *osmosis* could use the differing salinity of sea-water and fresh water to produce energy.

Power stations in the form of *floating islands* can produce energy using the *difference between the cold waters of the deep sea and the warmth of upper levels*, just as geothermal sources at present supply power on land.

The *power of ocean currents* can be harnessed by creating *under-water 'windmills'*. Double rotor blades would supply energy to coastal towns. The 'Seaflow' installation in the Bristol Channel, developed by the University of Kassel, is a first step.

Cities on piles: houses built on stilts in the water go back to the Bronze Age, and in South-East Asia millions of people still live in this way. The Japanese architect Kiyonori Kikutake has built a prototype village off Okinawa, and has plans to create cities of 50,000 to two million inhabitants, resting on an underwater mountain range off Borneo.

Growing islands: If you pass a direct current through an underwater wire lattice, deposits of calcium carbonate and magnesium hydroxide will form. The first island to be built up out of this 'bio-rock' is growing at the moment in nine metres of water above the Say-de-Malha sandbank between Madagascar and the Seychelles. A raft carrying solar cells serves to ionize the 'ecopolis'.

The inventor, architect and environmental designer Wolfgang Hilbertz originally developed these artificial reefs as a new habitat for threatened coral.

Fertilizing with iron: About three quarters of the oxygen given off on earth annually is produced by salt-water and fresh-water algae, especially the diatoms. But there are large areas of the ocean where the plankton population is very sparse, because the lack of iron in the water limits their growth. However, if you dress the sea with ferrous sulphate, for example, as scientists at the Alfred Wegener Institute have recently done in tests in Antarctic waters, plankton growth increases significantly. Why should you do this? Because the photosynthesis process traps and stores the greenhouse gas carbon dioxide as a compound. Researchers calculate that an ocean fertilized with iron could absorb more CO_2, and so reduce the greenhouse effect and its dangerous side effects.

Mining: There are valuable raw materials on the seabed: manganese nodules, for example, or massive sulphide deposits; gas hydrates; alluvial deposits; copper, zinc and gold – all just waiting to be mined.

Algae: They have a protein content of up to 70%, more than any other living creature. They are rich in iron (good), iodine (too much) and heavy metals (very bad). If the undesirable constituents of brown, red and green algae could be filtered out, nobody on earth would ever go hungry again.

Wave-powered generators

Articulated raft: a generator collects the energy produced by the movement between waves and mooring.

Nodding ducks: A famous device for converting oscillating mechanical energy to electrical energy. Professor Stephen Salter of the University of Edinburgh invented this string of heavy flapping objects resembling the cams on a camshaft (or ducks) which absorb energy from incoming waves and rotate about a 'spine' that leaves a calm sea in its lee.

Floating bellows: the Pelamis device. A long hinged rubber tube, the size of five railway carriages. Divided into separate chambers by flexible bulkheads, the tube moves up and down in the waves, and is pushed out of shape by their action. The chambers are compressed, and then expand again, so that as the tube bends at the hinges, they pump hydraulic fluid, which powers generators.

Wave ramp: a funnel-shaped ramp. As waves wash over it, the water is directed by the funnel into a basin situated above sea level. A low-pressure hydraulic power unit converts the run-off into energy.

Oscillating water column: comprises a partly submerged structure ('collector') which is open to the sea below the water surface, so that it contains a column of water. As waves enter and exit the collector, the water column moves up and down and acts like a piston on the air, pushing it back and forth. The air is channelled towards a turbine and forces it to turn. The turbine is coupled to a generator to produce electricity.

How many watts can the sea generate?

The sea works hard. The average tidal range of all the oceans put together produces 40,000 gigawatt hours, equivalent to all the energy generated per day throughout the world.

In Hamburg, up until the 19th century, tidal wheels pumped out the effluent of the city. From 1580 onwards similar machinery under London Bridge supplied fresh water. The device was in operation for 250 years.

Biggest tidal variation

Between high and low tides (not counting storm tides)
Fundy Bay, Canada . . . . 16.20 m
Granville Harbour, France . . . . 14.70 m
Severn Estuary, England . . . . 14.50 m
La Rance, France . . . . 13.50 m
Sea of Okhotsk, Russia . . . . 13.40 m
Puerto Rio Gallegos, Argentina . . . . 13.30 m
North Sea . . . . 3.00 m
Baltic Sea . . . . 30 cm

He'enalu: a very short history of surfing

Some time in the two millennia between 1500 BC and 400 AD Polynesians began paddling on planks through the surf: probably fishermen who had previously steered their canoes and outrigger boats skilfully through the waves. It is not clear when work became play and turned into one of the first sporting activities in human history. The first written account dates from 1777, when James Cook visited Tahiti. *He'enalu*, as the natives called their pastime, spread across the whole of the South Seas, from

New Zealand to Hawaii, and from the Easter Islands to New Guinea.

On other coasts too, in Peru and West Africa, Europeans were amazed to see youngsters throwing themselves into the sea clutching a board. In Hawaii it even became the sport of kings, hemmed round with dances, songs, rituals and taboos that kept the common people at a distance. When traditional culture in Hawaii and Polynesia was displaced by modernity, *He'enalu* also withered away. At the beginning of the 20th century in Waikiki and Oahu surfing had fallen out of fashion; Kona, the former capital of the sport, went into decline. Princess Kaiulani was the last traditional surfer to glide through the surf on her *wili wile ola* (oversized lightwood surfboard). Robert Louis Stevenson saw her in 1899, and wrote a poem about her.

In the first decade of the following century, the surfing scene was revived by schoolboys, the 'Waikiki Beach Boys'. The most famous of them was George Freeth, who as a lifeguard rescued 78 people from the sea on his surfboard. The new boards, at two metres long, were lighter and smaller than the traditional planks, and the first surfing club, the Outrigger Canoe Club, was dominated by whites; the original Hawaiians met in the *Hui Nalu*, among them the legendary Duke Kahanamoku, a gold medallist in the 100-metres freestyle swimming event at the 1912 and 1920 Olympics. When he went to Australia for a swimming competition in 1915, he built himself a surfboard and showed the spectators how to stand on it. A friend of Duke's, Tom Blanke, introduced the sport to California. From 1930 onwards, surfing became a competitive sport, and in the 50s the first surfing equipment suppliers began making boards from balsa wood. Within a few years, an exotic pastime had become a multimillion dollar business with its own sub-culture. What may be the oldest sport in the world has spread through all the oceans.

World champion surfers

7 times Kelly Slater 1992, 1994–8, 2005
6 times Layne Beachley 1998–2003
4 times Mark Richards 1979–82
4 times Freida Zamba 1984–6, 1988
4 times Lisa Anderson 1994–7
4 times Wendy Botha 1987, 1989, 1991, 1992

Surf Talk (a Benny's guide)

'woodies' . . . . estate car
(station wagon) with wooden
cladding on the sides
'polys', 'sticks' . . surfboards
'landshark' . pseudo-surfer
'soup' . . . . surf
'hairy' . . . . difficult (wave)
'filthy' . . . . excellent
'mamboosah' . . . . powerful

'shoot' . . . . a ride on the board
'hot dogging' . . . . surfing tricks
'bunnies' . . . . girls
'grom' . . . . surfer under 14
'troll' . . . . esoteric surf freak
'gremlins', 'kooks' . . . beginners,
idiots
'benny' . . . . pale tourist

Surfboard classifications in Hawaii

Olo – up to 100 kilos in weight, and seven metres long. Difficult to
steer, only suitable for a few waves. Not much in use since the
1930s. Olos made with light *wili wili* wood were reserved for the
chiefs; ordinary Hawaiians had to be content with the heavier *koas*.

Kiko'o – four to six metres long. For rough waves. Fast, but not particularly manoeuvrable.

Alaia – two to four metres long, about 40 centimetres wide, and
one to three centimetres in thickness; for centuries this was the
classic board for young and old.

Paipo – short and compact, it is reminiscent of present-day 'boogie
boards'.

The best surfing films

Gidget, 1959 ≈ *Big Wednesday*, 1978
The Endless Summer, 1964 ≈ *Five Summers Stories*, 1972
Point Break, 1991 ≈ *Apocalypse Now*, 1979
Blue Crush, 2002

Sea music III

Surf

Real surf rock is instrumental; the guitar is picked hard and fast, with no minor-key chords. In 1961 Dick Dale and his Deltones played at the Balboa Club in southern California, and the surfers present were enthusiastic. Dale became the guitarist idol of *Endless Summer*. More and more bands were formed, and some began to sing with the music: the Beach Boys, and Bruce and Terry. It was Terry Melcher, Doris Day's son, who introduced a crazed hippy into the scene, keen to join in the music-making. He wrote a song for the B side of a Beach Boys single ('Never learned not to love'). His name is Charles Manson, and he has been in jail since 1969 for his part in a spate of murders, including that of Sharon Tate, wife of the film director Roman Polanski.

Classic Surf hits (instrumental)

'Wipe out', *The Surfaris*, 1963 ≈ 'Miserlou', *Dick Dale*, 1962
'Mr Moto', *Bel-Airs*, 1961 ≈ 'Pipeline', *The Chanteys*, 1963
'The Pyramid Stomps', *The Pyramids*, 1964
'Surf Rider', *The Lively Ones*, 1963 ≈ 'Steel Pier', *The Impacts*, 1963
'Baja', *The Astronauts*, 1963

Surf hits (vocal)

'Surfin', *Beach Boys*, 1961 * 'Linda', *Jan and Dean*, 1963
'Summer means fun', *Bruce and Terry*, 1964

a different view of the world (1) – Zetetics

Why does a cannonball shot vertically into the air fall back down towards the cannon, if it is true that the planet is moving? Doesn't this phenomenon prove that the earth stands still and is a flat disc? The British Prime Minister Lord Palmerston was so bemused by this argument used by the zeteticists, more commonly known as 'flat-earthers', that in 1857 he ordered his War Secretary to conduct experiments to see whether more care should not be given to aiming a cannonade than hitherto. (Not that bullets missed their targets when fired simply because the earth was moving.)

Another example: If the earth's surface is curved, why is it possible to see the light of a seven-metre-high lighthouse (such as the light on Ryde Pier on the Isle of Wight) from the deck of a three-metre-high ship at a distance of 22.5 kilometres? According to orthodox opinion, from a height of three metres the horizon would be 6.5 kilometres away. That should mean that, seen from the ship, and subtracting the height of the lighthouse, the source of light would be 22 metres below the horizon.

This calculation, made by
Samuel Birley Rowbotham in his
book *Zetetic Astronomy: The Earth is not a
Globe* (1865), is mathematically correct. From this it
follows, in the view of believers in zetetics (from the
Greek *zētētikós*, 'inclining to examine, to research'), that the
earth is disc-shaped, just as the Bible says it is. The moon is
transparent and has its own light, and the sun stands at a fixed
point over our heads. In the middle of the earth an iceberg projects,
the North Pole, and the edge of the world is closed off by ice, the
Antarctic. The zeteticists regarded it as powerful proof of their theory
that no one was able to surmount this barrier and arrive at the edge of
the world. Although Roald Amundsen reached the so-called South
Pole on 14 December 1911, that told us nothing about how things
might look on the other side. When Vivian Ernest Fuchs did
succeed in crossing the frozen continent in 1958, the zeteticists
were not daunted. Previously they had known the earth was
not round; now, they simply *believed* it. The flat-earthers
have notched up one success, however: their conception of the world, a flat disc with the North Pole
in the middle, still features on the blue
United Nations flag.

The America's Cup yacht race

This cup is the oldest active trophy in international sport. The first
regatta took place on 22 August 1851 with the 28.5 m schooner yacht
America (New York Yacht Club Syndicate) beating 15 Royal Yacht
Squadron yachts around the Isle of Wight by 20 minutes. When
Queen Victoria asked who was second, the immortal answer was:
'There is no second, your Majesty.'

The United States held the trophy for the next 132 years until,
in 1983, the Australian challenger *Australia II* ended the longest

winning streak in sporting history. Since 1992 the competing yachts must conform to the International America's Cup Class rules, which require a monohull sloop of 24.4 m length.

In 2007 the Swiss *Alinghi* team will defend the Cup at Valencia, Spain – the first time for 150 years that the regatta has been held in Europe.

Guano

In 1804 Alexander von Humboldt brought some guano back from South America, and wrote to his chemist friend Martin Klaproth: 'The word *huanu* (Europeans are always mixing up *hua* with *gua* and *u* with *o*) in the Inca language means dung used for fertilizer. The verb 'to fertilize' is *huanuchani*. The aboriginal natives of Peru all believe that guano is formed from bird-droppings; but many of the Spaniards doubt it.' From 1845 onwards guano, rich in nitrates, became a leading fertilizer, and was later also used to make explosives. (It is well known that a primitive explosive can be made using artificial fertilizer and sugar. Don't try this at home.)

Saltpetre chaos

After the m on k Berthold Schwarz invented gunpowder, there was an incr eased demand for saltpetre or nitre. The travelling saltp etre collectors went rooting around in farmyards for the valuable mineral, which accumul ated under stables an d dung heaps. Cou ntry people regarded the salt petre-hunters as pests, since they received no com pensation for the chaos they left behind. An other way of ob taining saltpetre was to ref ine it out of eel grass or kelp, a method first tested by the Fren chman Curtois in 1811. As a by-product of this pro cess another substance was gained, sup erfluous to sa ltpetre prod uction – iodine. It was the first time it had been obtained in a chem ically pure form. This finding was made just in time, beca use kelp became obsolete as a raw material once the great sa ltpetre deposits in the Chil ean Atac ama Desert we re disco vered, and the tra de of saltpetre-refiner died out soon afterwards.

Insurance fraud (1)

London 1783: James Gregson, part-owner of the *Zong*, demanded payment from his insurance company for the cargo he had lost off Jamaica. The goods had gone bad, and the crew had therefore thrown them overboard for the safety of the ship. But the court was able to prove that there were still 420 gallons of water stored on board, so that the cargo could not have been ruined. His claim was dismissed.

The 'goods' concerned were slaves. Many had become sick through overcrowding. Chained together, they had less space than in a coffin. Sixty had already died. ('For the safety of the ship' implied, untruthfully, that there was not enough water to keep the slaves alive for the rest of the journey.) The fact was that sick slaves were worthless, but the insurers would not pay up if a slave died on board, since the insurance company deemed it be to the result of poor 'cargo management'. They would only pay if a slave went over the side alive. So Captain Luke Collingwood had thrown 133 Africans overboard. It is said that the last ten defiantly leapt into the sea, triumphantly embracing death. Gregson's case against his insurers was rejected. But no one was ever held to account for the murder of the Africans.

A different view of the world (2): Cosmic Ice Theory

Once upon a time in the Columba constellation a massive body of ice collided with a powerful sun. The ice formed a metallic crust, which is why it did not immediately evaporate, but was contained for a while in this glowing womb. Eventually, it caused a great explosion, and the water condensed out into ice, forming a ring that we now call the Milky Way. The outer planets Mars, Jupiter, Saturn, Uranus and Neptune are blocks of ice, and so is our moon. When ice comets collide with the sun, the result is sun spots. Fine space dust is still falling to Earth; we call it hail.

The Austrian engineer Hanns Hörbiger – incidentally also known for the invention of the plate compressor valve – had this blinding revelation when observing the moon, and devoted his life to his 'Glacial Cosmology', supported by the astronomer Philipp Fauth. Even after Hörbiger's death in 1931, his WEL (Welteislehre, or Cosmic Ice Theory) still had faithful adherents, particularly among Nazis. Goebbels even felt obliged to proclaim that 'one can be a good National Socialist without believing in the WEL'. One of the Nazi believers was the leader of the SS Ancestral Heritage department; his name was Heinrich Himmler. Under the rational-sounding heading of meteorology, he tried to prove Hörbiger's theory. Himmler ordered an archival record to be kept of hailstorms – i.e. cosmic dust – and the counting of sand ridges on the seashore in Ethiopia. He also appointed Philipp Fauth to head the observatory at the German Museum in Munich. Fauth, still a respected name in moon research, called a crater after Hörbiger. It was later renamed after Fauth.

Sea-level anomalies (1): trouble with zero surface

Water has the nice property of distributing itself evenly, which is why it was practical to use sea level (or 'zero surface') as a given standard by which to measure elevation. But in practice, there were discrepancies, since the sea is not static. Sea levels vary with the tides, the wind causes waves, there are different factors in summer and winter. So 'mean sea level' (MSL) averages out all these effects to a single 'mean' level. Mean Sea Level is defined as the hypothetical point of interface between air and the surface of an imaginary 'global ocean'. This too can vary; for example, the MSL on the Pacific Coast of North America ranges from 8 to 12 metres higher than on the Atlantic Coast. Standards have had to be established, e.g. the Amsterdam Tide Gauge, a bronze bolt in a concrete pillar that can be seen in the Stadthuis (city hall) in Amsterdam.

Bremerhaven. It seemed like a normal December day in 1875 when suddenly, as the steamer *Mosel* was being unloaded, a barrel exploded and blew up the ship together with a tug tied up alongside it, the *Simon*. 81 people were killed, and over 200 injured. During the investigation, a percussion device and explosives were found. Suspicion soon fell upon a certain William King Thomas, who had insured the cargo for a large sum. Born Alexander Keith junior, the red-haired Scotsman had previously made ends meet as a beer-brewer in Canada, black marketeer and smuggler during the American Civil War, and insurance fraudster in Germany. Accused of the crime, he shot himself in the head. He lived for a few days before succumbing to his injuries. His head was removed from the body in order to examine it for possible illnesses (this was in the days when it was thought possible to identify a born criminal from the shape of his head). The head was preserved in formaldehyde in the Natural History Museum in Bremerhaven until 1944, when the building was destroyed by another explosion – an air raid this time.

Sea quotations

'Here he lies where he longed to be;
Home is the sailor, home from the sea,
And the hunter home from the hill.'
Robert Louis Stevenson (1850–1894), from 'Requiem' 1887

'The sea is the universal sewer'
Jacques Cousteau (1910–97), in 1971 to the House Committee on Science

'Where seagulls follow a trawler, it is because they think sardines will be thrown into the sea.'
Eric Cantona footballer (1966–), Press Conference 1995

'The sea hates a coward!'
Eugene O'Neill (1888–1953), *Mourning becomes Electra* 1931

'Full fathom five thy father lies;
Of his bones are coral made:
Those are pearls that were his eyes:
Nothing of him that doth fade,
But doth suffer a sea-change
Into something rich and strange.'
William Shakespeare (1564-1616) *The Tempest* 1611

'I have seen the sea lashed into fury and tossed into spray
its grandeur moves the soul of the dullest man; but I
remember that it is not the billows, but the calm level of the
sea from which all heights and depths are measured.'
James Abram Garfield (1831–81), speech nominating John
Sherman for president 1880

'By the deep sea, and music in its roar:
I love not man less but nature more.'
Lord Byron (1788–1824), 'Childe Harold's Pilgrimage' 1812–18

'God moves in a mysterious way
His wonders to perform;
He plants his footsteps in the sea,
And rides upon the storm.'
William Cowper (1731–1800),
'Light Shining out of Darkness' 1779

10 Favourite sea poems

(Source: 'Sea Britain 2005')

| 1 | 'Sea Fever' | John Masefield | 1878–1967 |
| 2 | 'Song of the Waterlily' | Martin Newell | 1953 |

'The bird-shit law'

Uninhabited islands where a large amount of bird excrement is found automatically become American territory whenever an American citizen discovers them. The 'Guano Island Act', passed in 1856 to secure deposits of this important raw material for the USA, is in theory still in force today. The following islands have thus become American: Baker Island, Jarvis Island, Howland Island, Kingman Reef, Johnston Atoll and the Midway Atoll. A few countries have raised objections: Navassa Island is claimed by Haiti, and the Serranilla Bank by Nicaragua. No fewer than three nations have registered claims to the Bajo Nuevo Bank – Jamaica, Columbia and Honduras.

Gold

Sea water contains 0.0000044 grams of gold per tonne of water (which works out at 777 million tonnes altogether). Rocks on land with gold-bearing lodes may contain 4 grams per tonne. On the Conical Seamount (at a depth of 1600 metres) near Papua New Guinea the concentration is 13 grams per tonne, and experts estimate that the rocks of the Mermadec subduction zone near Tonga (at a depth of 1700 metres) contain as much as 30 grams per tonne. But at this depth, how can it be got at?

As small children we were taught that the Earth is a sphere. That is why, when a ship appears above the horizon, you see first its highest point, and then the hull, as though the ship were coming up from below.

But that is an optical illusion. Beyond the horizon, the earth curves upwards. We are not living on the outside of the globe, we are crawling around on the inside of a hollow sphere, like flies in a glass. In this hollow sphere, with a diameter of 12,000 kilometres, there is room for the whole cosmos. In the centre, less than 6,000 kilometres from the Earth's surface, a spherical fixed star turns. Out of its pores, as though out of a Swiss cheese, shines the night sky as we know it. What seems to be millions of light years away, from Proxima Centauri to the Sefert galaxies, are actually sparks of light from this disco-ball. Our solar system revolves around this glittering core: the sun, the planets including Earth, and the moon. The sun, planets and moon are obviously much smaller than is generally assumed. The moon only measures about 200 kilometres in diameter, by some calculations only a single kilometre, and not 3476 km, as orthodox astronomers believe.

What causes day and night, then? Why can a plane not take a short cut and reach Australia in a few hours? Why can't we see the lights of New York at night above the horizon?

The answer is simple: light does not travel in a straight line, but is curvilinear. The speed of light does not spread constantly. Nobody has yet measured the speed of light in space; it has only been estimated. For distances in the cosmos are calculated through angles, and not measured by length. Gravitation may have slowed the astronauts down and forced them into a spiral trajectory.

'You can reflect the whole world in its interior,' claimed Professor Rudolf Kippenhahn. 'It's called reflection through reciprocal radii.' Professor Kippenhahn was Director of the Max Planck Institute for Astrophysics in Garchen near Munich; a proper, serious professor of astronomy. 'In the centre the distant stars are piled up. Accordingly, the moon has a diameter of only one kilometre. Straight lines become arcs of a circle which run through the centre.'

The Austrian physicist Roman Sexl (1939–86) helped the hollow-Earth theorizers to construct an internally consistent, irrefutable cosmography. Similarly to the theory of relativity, where time is reduced the closer to the speed of light one comes, the further astronauts fly in space the smaller they become. On the moon they would have been mouse-sized and weighed 20 grams. 'In the hollow world as it is according to Sexl, you've got to re-calculate all the familiar laws of nature. Appearances remain the same. Everything becomes endlessly complicated. Just think of centrifugal force. Sexl's theory contains no errors, and cannot be refuted, any more than it can be proved. Actually, it's completely unimportant whether we imagine we're inside or outside. We can never perceive the truth.' The pictures of Earth as a blue ball floating in space, the red shift, can all be explained by the curvature of light. We are trapped in a bubble; infinity is beneath our feet, not above our heads. This conception of an inner world is first and foremost a literary one: Jules Verne described in *A Journey to the Centre of the Earth* an expedition to the depths. After the participants have successfully passed through the zone of great heat, they enter a strange realm on the other side of the Earth's crust. In the 1920s this alternative vision of the world had millions of adherents. After the shock of the Sputnik, however, there were few left.

And yet the hollow globe theory has the enormous advantage of being irrefutable. If you accept that space is not homogeneous and not isotropic, so that physical properties such as the speed of light are variable, then the traditional Copernican view of the world can be turned inside out like a sock. The honeycomb-shaped universe is no naive analogy on the part of esoteric thinkers. For decades mathematicians and physicist have been searching for what Einstein called a Unified Theory. For Einstein's theory of relativity and Heisenberg's quantum mechanics are irreconcilable. The link is missing. Or perhaps there is something here that is simply superfluous? The British mathematician Julian Barbour cut out a troublesome element from a world theory developed by Wheeler and DeWitt, and

there it was – the perfect world formula. So beautiful, and so simple. As long as you ignore the fact that time would have to be eliminated, as a factor that spoils the theory. Seen mathematically, time ought not to exist, and then physics would be as unified and self-contained as in Newton's day. That is precisely what the Briton Julian Barbour postulates. There is no such thing as time. We live in time capsules, in which there is only the present. Yesterday is a different time capsule from today. By this reckoning, one time cosmos would stand next to another, in honeycomb formation, and new cells would be growing all the time. Everybody is familiar with time capsules. You, dear reader, have one in your hand at this minute. Just reach out, turn the pages of the book next to you, and Madame Bovary is falling in love all over again. Huckleberry Finn is drifting once more down the Mississippi …

Sea-level anomalies (2)

≈ The sea level in the Gulf of Corinth is seven metres higher than in the port of Patras.
≈ On the south side of Crete and Rhodes the water is 20 metres deeper than on the north side of each island.
≈ In the Indian Ocean south of Sri Lanka the sea curves to 105 metres above mean sea level (MSL), while in the region of the Nicobar Islands, in the same ocean, the curve measures only 40 metres, representing a drop of 3.9 centimetres per kilometre.
≈ The 'curvature' of the sea is caused by anomalies in the Earth's field of gravity, i.e. deviations in gravity which may be due to a mountain range beneath the sea, or the particular density of the rock mass. Such an unusual gravitational field attracts an unusual mass of water.

Insurance fraud (3)

In 1977 the *Lucona* sank off the Maldives. The cause was an explosion that killed the entire crew of six. The ship was allegedly transporting a uranium ore reprocessing plant worth 212 million Austrian Schillings. That, at least, was what it said on the insurance policy that Udo Proksch produced after the disaster. Proksch was a prominent Viennese socialite, with good contacts to the KGB and excellent connections in Austrian government circles. A committee of enquiry was set up to investigate the involvement of politicians: in the end, the affair brought down the President of Austria. It was established in court in 1992 that Proksch had arranged for the *Lucona* to be sunk by a time-bomb. For the six murders he was given a life sentence, and died in prison in 2001. (An interesting sidelight is that Proksch was – briefly – related by marriage to the Wagner family, of Bayreuth opera fame, and had attempted to blackmail them over supposed financial irregularities in 1969.)

Tide gauges: sea-level anomalies (3)

Country	Gauge	MSL minus Height of Land (metres)
Belgium	Ostend	− 2,30
Denmark	Copenhagen	− 0,09
France	Marseilles	− 0,25
Luxembourg	Amsterdam	0,00
Netherlands	Amsterdam	+ 0,02
Austria	Triest	− 0,25
Switzerland	Geneva	− 0,06

Alchemy

In 1910 Fritz Haber developed the process named after him for obtaining ammonia from hydrogen and nitrogen. The artificial fertilizer thus produced was acclaimed as 'bread from the air', and made expensive imported guano superfluous: the suppliers went out of business. But using the Haber-Bosch process you could also make 'explosives out of air'. Fritz Haber headed poison gas production in the First World War. He personally oversaw the deployment of chlorine gas at Ypres (3,000 died, 7,000 suffered burns). This led his wife to commit suicide with his service revolver. After the war the victorious allies demanded that Haber should be handed over as a war criminal. But then, in 1918, came a different kind of bombshell: Haber was awarded the Nobel Prize for chemistry. The Prize Committee made up for the war-time hiatus by celebrating the nine winners whose awards had been announced during the war. Among them were five Germans, and in the official 1920 photo can be seen, in their frock coats and bow ties, Max von Laue, Richard Willstätter, Max Planck, Johannes Starck and Fritz Haber, with four wives sitting very upright on the sofa in front of them. Clara Haber is missing.

To help pay the reparations demanded of Germany after the war, as head of the Kaiser Wilhelm Institute for physical chemistry in Berlin Haber tried for six years to obtain gold from seawater, in the style of the founding fathers of chemistry, the alchemists. On research trips with the *Hansa* and the *Württemberg* he and his colleagues took over 5,000 samples of seawater, which they filtered and centrifuged. The early analyses yielded such a high gold content that commercial development seemed viable. But plankton, silver, and the gold that had been melted down in a room next to the laboratory had falsified the results. The gold content was reduced with every improvement in the accuracy of testing procedures. Haber succeeded in demonstrating a concentration of one-billionth of a gram of gold per litre, but in the end he stood empty-handed. To produce a return, the concentration would have needed to be a thousand times stronger. His estimate of the amount of gold in the sea, however, still holds good.

In 1933, because of his Jewish origins (Haber had converted to Protestant Christianity in 1893), Haber was forced to resign. A few months later, he died while in the process of emigrating. Today the Chemistry Institute of the Max Planck Society is named after him.

Flettner's Rotor Ship

If the wind blows perpendicular to a static upright cylinder – a chimney, for example – the air current splits to both sides of the cylinder, flowing evenly. But if the cylinder is rotating, one side moves with the air current, and the other offers resistance to it. The result is that on one side of the cylinder the air is helped along by the rotation, and flows faster. On the other side the air is slowed down.

As with an aircraft wing, suction is created. On the side where the air moves faster, downward pressure builds up; on the slower side, upward pressure. As a result a strong downward pull is exerted upon the rotating cylinder. The pressure differential between the two sides results in a force pushing in one direction.

When the engineer Anton Flettner (1885–1961) worked the idea out on paper, there were some startling findings. Under optimal conditions – with a side wind, if the rotors were turning four times as fast as the wind speed – a cylinder should be able to develop 14 times as much forward thrust as a sail of the same size. In late July 1923 the inventor applied for a patent on his 'Flettner Rotor'. Shortly afterwards he launched a model ship on the Wannsee Lake in Berlin, fitted with paper cylinders fifty centimetres high that were rotated by clockwork. Flettner noted: 'The ship set off at a brisk speed, so that I immediately found my theoretical assumptions confirmed.'

In 1924 he managed to persuade the directors of the Germania Shipyard in Kiel to let him conduct trials on a full-scale model. They made over to him the small merchant ship *Buckau* to be fitted with rotors. By the end of September the 'Long Henry' crane was already placing two white-painted, 16-metre tall cylinders on the deck of the *Buckau*. On 4 October Chief Engineer Heinrich took the ship out for a first trial run. He gradually increased the turning speed of the rotors, to 20, then 50, and finally 60 revolutions a minute. In a light southwesterly breeze the strange ship got gently under way.

The official presentation on 7 November 1924 was a triumph. Both sides of the *Buckau* were boldly marked in white letters with the inscription 'Flettner Rotor' – evidence of a certain vanity on the part of the inventor. Local celebrities and visiting experts and reporters alike were astonished. The newsreel cameras of the UFA film company sent the news of the rotor ship around the world. Not a few of the euphoric spectators were already imagining whole fleets of rotor ships crossing the world's oceans. But the time was not yet ripe for the idea. Fuel was still too cheap, and Flettner's brilliant idea was forgotten. Jacques Cousteau tried to revive it in the early 1980s with his 'turbosail' system, intended to save fuel and cut pollution. But once again, the concept was never widely adopted. Perhaps its time is still to come.

Mercury

- ≈ 0.0001 milligrams of mercury per litre in the North Sea.
- ≈ 0.1 milligram per kilo of fish.
- ≈ 1 milligram per kilo in long-lived species such as shark, tuna and halibut.
- ≈ According to EU/European Commission regulations, adopted into UK law, 1 milligram per kilo is the maximum permitted amount for these species that naturally accumulate higher levels of mercury, such as eels, sturgeon, salmon, blue ling, Greenland shark, bottle-nose shark, Norway haddock, tuna or swordfish.
- ≈ 200 milligrams per kilo in the livers of Polar bears and seals.
- ≈ 0.3 milligrams per day can cause poisoning.
- ≈ The average ingestion of mercury per person in the UK from common fish (excluding species with high mercury levels) is approximately 0.1 milligrams per week.

Silicon

After oxygen, silicon is the commonest element on the planet. This metalloid represents 15% of the Earth by weight, rising to 26% for the Earth's crust. Billions of tonnes of silicates are leached out annually, dissolved in rivers, and carried away to the oceans, where they are held in a compound by the diatoms (single-cell algae), which absorb silica into their skeletons. As the algae also take up CO_2, they have acquired an important role on the global climate scene. When the minute plankton die, they take with them down to the seabed the carbon dioxide they have absorbed, thus reducing the level of greenhouse gas in the atmosphere. But the natural circulation of silicon is now being increasingly interrupted by the building of dams, with the result that supplies of silicon in the seas are reduced, which in turn means a reduction in the quantity of algae and their capacity to trap CO_2.

Sportsmen's nicknames

Shark ≈ Greg Norman, Golf
Shrimp ≈ Roy Worters, Ice hockey
Squid ≈ Sidney Moncrief, Basketball
Walrus ≈ Craig Stadler, Golf

Walking round the world

Karl Bushby, a 37-year-old British ex-paratrooper, aims to be the first person to walk non-stop across all the world's continents. He started on 1 November 1998 from Punta Arenas in Chile. At the time of writing he had completed 27,000 kilometres, with 30,000 still to go. In March 2006 he crossed the frozen Bering Strait from Alaska, a distance of 90 kilometres, and ran into great difficulties on the Russian side: he was arrested for violation of border crossing rules. The British Deputy Prime Minister appealed to the billionaire Roman Abramovich, owner of Chelsea Football Club and governor of the autonomous region of Chukotka. His successful intervention allowed Karl Bushby to continue his walk. Early in 2007, Karl Bushby ran into further visa difficulties, and at the end of May he was still waiting to continue his walk across Russia, hoping to return to Chukotka from Alaska in late July 2007. He will then have nineteen thousand more miles to go. When he reaches the English Channel he hopes to receive special permission to walk through the Channel Tunnel. At the present rate of progress he is expected to return to Hull in 2012.

How to get to St Helena

St Helena is a tiny island in the South Atlantic, with no natural harbour and no airport. You can get there on the RMS *St Helena* either from Tenerife via Ascension Island (14 days), or else from Cape Town via Walvis Bay (7 days). (It took Napoleon ten weeks to reach his final place of exile after his defeat at Waterloo.) The voyage today from Tenerife to St Helena on the last British Royal Mail Ship costs from £567. For timetables, see www.rms-st-helena. com/voyageschedules.html

Maritime myth?

Supposedly genuine radio exchange between Galicians and Americans on October 16 1997:

Galician: A-853 speaking, please change course 15 degrees south-wards to avoid a collision. You are heading directly for us, distance 25 sea miles.

American: We advise *you* to change course 15 degrees northwards to avoid a collision.

Galician: Negative. We repeat: change your course by 15 degrees to south to avoid a collision.

American (a different voice): This is the captain of a United States Navy ship speaking. Change your course immediately by 15 degrees to north to avoid a collision.

Galician: We do not think that is either possible or necessary: we advise you to change 15 degrees southwards to avoid a collision.

American (agitated, emphatic tone of command): Captain Richard James Howard speaking, commander of the aircraft carrier USS *Lincoln* of the United States Navy. This is the second largest bat-tleship of the North American fleet; we are being escorted by two battlecruisers, six destroyers, five cruisers, four submarines and a number of other ships ready to support us at any time. We are heading for the Persian Gulf for manoeuvres. I don't advise you, I *order* you to change course 15 degree northwards!

Galician: Juan Manuel Salas Alcántara speaking. We are two people. We are escorted by our dog, our food, two beers and a canary, currently asleep. We have support from the transmitters Cadeña Dial of La Coruña, and Channel 106 for putting out maritime distress signals. We are not heading anywhere, since we are situated in lighthouse Finisterre A-853 on the Galician coast.

Islands on euro notes

The three dots above the serial number are the Azores, with Madeira below them. Right at the bottom, in the boxes next to the Greek word EYPO, are the French overseas *départements* of French Guyana: Guadeloupe (at the top), Martinique (centre) and Réunion (below). Next to them are the Canaries, though Gomera and Hiero are missing, and neither Cyprus nor Malta is represented.

Submarines lost

In World War II:

Germany	784	USA	52
Japan	112	Netherlands	11
USSR	109	Poland	2
Italy	82	Norway	1
Great Britain	75		

And since:

USA:

Scorpion (SSN-589)	27 May 1968
Thresher (SSN-593)	10 April 1963
Cochino (SS-345)	26 August 1949

The sea, according to the US coast guard's lifesaving manual

'The ocean is always looking for a way into your boat'

Man overboard

≈ If you fall into the water, you should move as little as possible, in order to conserve your body heat. Legs pulled up, arms folded, back bent: this posture will reduce your total surface area and thus your heat loss.

≈ Do not swim. Let yourself drift.

≈ Keep clothing on. Wool is best because it keeps in warmth even when wet.

≈ Try to stay in groups and tie yourselves together with a rope.

≈ Protect your nose and mouth from splashing water. Salt water in the lungs can be fatal.

Index

227

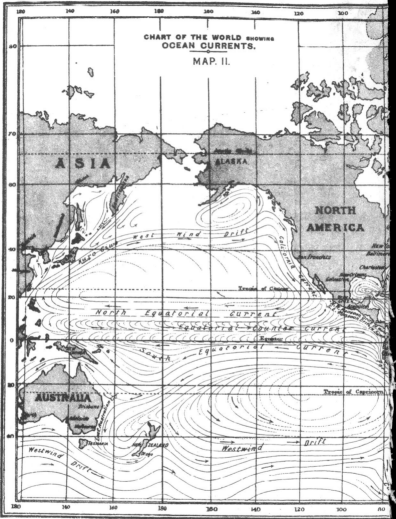

CHART OF THE WORLD SHOWING
OCEAN CURRENTS.
MAP. II.